GRAVITATIONAL WAVE ASTROPHYSICS: EARLY RESULTS FROM GRAVITATIONAL WAVE SEARCHES AND ELECTROMAGNETIC COUNTERPARTS

IAU SYMPOSIUM 338

COVER ILLUSTRATION:

Illustration of the merger of two black holes producing gravitational waves.

Photo Credit: LIGO/T. Pyle

IAU SYMPOSIUM PROCEEDINGS SERIES

INTERNATIONAL ASTRONOMICAL UNION
UNION ASTRONOMIQUE INTERNATIONALE

International Astronomical Union

IAU

GRAVITATIONAL WAVE ASTROPHYSICS: EARLY RESULTS FROM GRAVITATIONAL WAVE SEARCHES AND ELECTROMAGNETIC COUNTERPARTS

PROCEEDINGS OF THE 338th SYMPOSIUM OF THE INTERNATIONAL ASTRONOMICAL UNION HELD IN BATON ROUGUE, UNITED STATES
OCTOBER 16-19, 2017

Edited by

GABRIELA GONZÁLEZ
Louisiana State University, United States

and

ROBERT HYNES
Louisiana State University, United States

CAMBRIDGE
UNIVERSITY PRESS

CAMBRIDGE UNIVERSITY PRESS
University Printing House, Cambridge CB2 8BS, United Kingdom
1 Liberty Plaza, Floor 20, New York, NY 10006, USA
10 Stamford Road, Oakleigh, Melbourne 3166, Australia

First published 2019

Printed in the UK by Bell & Bain, Glasgow, UK

Typeset in System LaTeX 2ε

A catalogue record for this book is available from the British Library Library of Congress Cataloguing in Publication data

ISBN 9781107192591 hardback
ISSN 1743-9213

Table of Contents

Preface vii

Editors viii

Participants ix

Revealing Short GRB Jet Structure and Dynamics with Gravitational Wave Electromagnetic Counterparts 1
Gavin P. Lamb & Shiho Kobayashi

Gravitational wave optical counterpart searching based on GRAWITA and DLT40 project during LIGO O2 run 9
Sheng Yang & on behalf of GRAWITA and DLT40 project

On the host galaxy properties of stellar binary black hole mergers 14
Youjun Lu, Liang Cao & Yuetong Zhao

Reanalysis of LIGO black-hole coalescences with alternative prior assumptions 22
Davide Gerosa, Salvatore Vitale, Carl-Johan Haster, Katerina Chatziioannou & Aaron Zimmerman

High accuracy measurement of gravitational wave back-reaction in the OJ287 black hole binary 29
Mauri J. Valtonen, L. Dey, R. Hudec, S. Zola, A. Gopakumar, S. Mikkola, S. Ciprini, K. Matsumoto, K. Sadakane, M. Kidger, K. Gazeas, K. Nilsson, A. Berdyugin, V. Piirola, H. Jermak, K.S. Baliyan, D. E. Reichart, D. B. Caton, S. Haque & the OJ287-15/16 Collaboration

Background Rejection using Convolutional Neural Networks. 37
Adam Zadrożny & Beata Goźlińska

Merging massive black holes the right place and the right time. 40
Astrid Lamberts

Supermassive Black Hole Binary Candidates from the Pan-STARRS1 Medium Deep Survey 46
Tingting Liu & Suvi Gezari

GW170817: *Swift* UV detection of a blue kilonova, and improving the search in O3 53
Aaron Tohuvavohu, Jamie A. Kennea & the Swift *gravitational wave follow-up group*

DESGW Optical Follow-up of BBH LIGO-Virgo Events with DECam 61
Robert E. Butler, M. Soares-Santos, J. Annis, K. Herner & for the Dark Energy Survey Gravitational Wave program and the DECam Community Group Collaboration

Cosmology with Gravitational Waves in DES and LSST 65
Kenneth Herner, Marcelle Soares-Santos, James Annis & on behalf of the Dark Energy Survey Gravitational Wave program and the DECam Community Group Collaboration

Observations of GW170817 by DESGW and the DECam GW-EM Collaboration 72
James Annis

Observations of the first electromagnetic counterpart to a gravitational wave source by the TOROS collaboration. 80
Lucas M. Macri, Mario C. Díaz, Diego Garcia Lambas & the TOROS *collaboration*

AGILE Observations of GW Events. 84
Francesco Verrecchia, Marco Tavani & Immacolata Donnarumma, on behalf of the AGILE Team

High power in Advanced LIGO. 90
Terra Hardwick & for the LIGO Scientific Collaboration

Strong-lensing of Gravitational Waves by Galaxy Clusters 98
Graham P. Smith, Christopher Berry, Matteo Bianconi, Will M. Farr, Mathilde Jauzac, Richard Massey, Johan Richard, Andrew Robertson, Keren Sharon, Alberto Vecchio & John Veitch

How Einstein's Theory of Relativity Gives us $E = mc^2$ and the Atomic Bomb 103
Richard Conn Henry

Author index 107

Preface

The detection of gravitational waves in September 2015 with the LIGO detectors opened a new era in gravitational-wave astronomy. The first detection and the ones that followed have provided insights into previously unanticipated black holes of stellar size. The detection of the merger of neutron stars with LIGO and Virgo detectors, and the many electromagnetic counterparts were announced on the first day of the workshop, started a new era of multi-messenger astronomy.

Physicists and engineers working on the sensitive detectors, and physicists and astronomers working on the analysis and interpretation of the data got together in Baton Rouge to share knowledge and enthusiasm for science with gravitational waves. The conference attracted a very diverse participation with 83 participants ranging from beginning graduate students to very senior figures in the field from many countries that exchanged ideas during three days of talks.

Taking advantage of impeccable, albeit coincidental, conference timing, the conference started watching live on the screen the announcement in Washington, DC, of the discovery and follow up of the merger of neutron stars dubbed GW170817; several of the expositors in the press conference flew to baton Rouge to give talks in the conference in the next days. The conference had talks on both Earth- and space-based detectors and detection of gravitational waves using pulsar timing, as well as sources of gravitational waves and searches for those including not just binary systems but also periodic signals from rotating stars, stochastic backgrounds from unresolved sources and cosmological sources, and un-modeled sources.

In our conference banquet, we remembered sadly the passing away of Dr. Neil Gehrels, a famous and honored astrophysicist who was at the time of his death the President of the IAU Commission D.1 on Gravitational Wave Astrophysics, and a co-chair of the Scientific Organizing Committee of this conference.

It was a very timely conference for a very exciting time that organizers and participants enjoyed thoroughly, thanks to the financial and logistics support of the International Astronomical Union and Louisiana State University.

Gabriela González and Robert Hynes

Editors

Gabriela González
Louisiana State University, United States

Robert Hynes
Louisiana State University, United States

Organizing committee

Scientific Organizing Committee

SOC Chair

Gabriela González	Louisiana State University, United States

SOC Members

Masaki Ando	University of Tokyo, Japan
Pierre Binétruy	University Paris Diderot, France
Marica Branchesi	Urbino University, Italy
Lynn Cominsky	Sonoma State University, United States
Neil Cornish	Montana State University, United States
Karsten Danzmann	Max Planck Institute for Gravitational Physics, Germany
Federico Ferrini	European Gravitational Observatory, Italy
Vicky Kalogera	Northwestern University, United States
Richard N. Manchester	CSIRO, Australia Telescope National Facility, Australia
David Reitze	California Institute of Technology, United States
David Shoemaker	Massachusetts Institute of Technology, United States
Tarun Souradeep	The Inter-University Center for Astronomy and Astrophysics, India
Stan Whitcomb	California Institute of Technology, United States

Local Organizing Committee

LOC Chair

Gabriela González	Louisiana State University, United States

LOC Members

Robert Hynes	Indian Institute of Astrophysics, India
Stephanie Jones	Special events coordinator
Mimi LaValle	Public Relations/Marketing
Yao Zeng	Website coordinator

Participants

Thomas	Abbott	Louisiana State University
Haakon	Andresen	Max Planck Institute for Astrophysics
James	Annis	Fermilab
Javier Mauricio Antelis	Ortiz	Technologico de Monterrey en Guadalajara
Iair	Arcavi	UC Santa Barbara/Las Cumbres Obs.
Stuart	Astin	LIGO Livingston Observatory
Corey	Austin	Louisiana State University
Martin	Beroiz	University of Texas Rio Grande Valley
Carl	Blair	Caltech/LIGO Livingston Observatory
Marica	Branchesi	GSSI
Robert	Butler	Fermilab/Indiana University
Elvis Camilo	Ferreira	INPE (Brazil)/LLO
Richard Salvatore	Camuccio	University of Texas Rio Grande Valley
Nicholas	Cannady	Louisiana State University
Moises	Castillo	Louisiana State University
Sourav	Chatterjee	CIERA-Northwestern University
Michael	Cherry	Louisiana State University
Min-A	Cho	University of Maryland
Lynn	Cominsky	Sonoma State University
Thomas	Corbitt	Louisiana State University
Ron	Cowen	freelance science writer
Jonathan	Cripe	Louisiana State University
Liang	Dai	Institute for Advanced Study
James	DeLauney	Penn State
Matteo	Di Giovanni	TIFPA
Susan	Dorsher	Louisiana State University
Maria	Drout	Carnegie Observatories
Anamaria	Effler	LIGO Livingston Observatory
Liudmila	Fesik	Saint Petersburg State University
Stanislav	Fisenko	JSC "Rusthermosynthesis"
Ryan	Foley	UC Santa Cruz
Juhan	Frank	Louisiana State University
Valery	Frolov	LIGO Livingston Observatory
Juan Leonel	Garcia	University of Texas Rio Grande Valley
Davide	Gerosa	Caltech
Joseph	Giaime	LIGO Livingston, Caltech/LSU
V. Zach	Golkhou	University of Washington
Gabriela	Gonzalez	Louisiana State University
Sandy	Graham	Orange County Astronomers
Scott	Graham	Orange County Astronomers
Rachel	Hamburg	University of Alabama in Huntsville
Terra	Hardwick	Louisiana State University
Jan	Harms	GSSI
Matthew	Heintze	LIGO Livingston Observatory
Richard	Henry	Johns Hopkins University
Kenneth	Herner	Fermilab
Andy	Howell	Las Cumbres Observatory/UCSB
Robert	Hynes	Louisiana State University
Kunihito	Ioka	Yukawa Institute for Theoretical Physics
Marie	Kasprzack	Louisiana State University
Jamie	Kennea	Penn State
Keiko	Kokeyama	KAGRA Observatory
Gavin	Lamb	Astrophysics Research Institute
Astrid	Lamberts	Caltech
Arlo	Landolt	Louisiana State University
Benjamin	Lane	Louisiana State University
Tyson	Littenberg	NASA Marshall Space Flight Center
Tingting	Liu	University of Maryland
Youjun	Lu	National Astronomical Observatories of China
Lucas	Macri	Texas A&M University
Cody	Messick	Penn State
Brian	Metzger	Columbia University
Felix	Mirabel	Institut of Astronomy and Space Physics
Claudia Moreno	Gonzalez	Universidad de Guadalajara
Patrick	Motl	Indiana University Kokomo
Adam	Mullavey	LIGO Livingston Observatory
Richard	Oram	LIGO Livingston Observatory
Brian	O'Reilly	Caltech - LIGO
Michael	Pajkos	Michigan State University
Gary	Pearson	Independent researcher
Arnaud	Pele	LIGO Livingston Observatory
Tania	Penuela	University of Texas Rio Grande Valley
Jorge	Pullin	Louisiana State University
Karla	Ramirez	University of Texas Rio Grande Valley
Jared	Rice	University of Nevada, Las Vegas
George	Robinson	Palmia Observatory (Retired)
Janeen	Romie	LIGO Livingston Observatory
Skye	Rosetti	University of Leicester
Eva	Sagi	Technion
Brad	Schaefer	Louisiana State University
Peter	Shawhan	University of Maryland
Karelle	Siellez	Georgia Institute of Technology
Jacob	Slutsky	NASA Goddard Space Flight Center
Graham	Smith	University of Birmingham
Marcelle	Soares-Santos	Brandeis University
Joel	Tohline	Louisiana State University
Aaron	Tohuvavohu	Penn State University
Maggie	Tse	LIGO MIT
Rhondale	Tso	Caltech

Guillermo	Valdes	Louisiana State University
Stefano	Valenti	UC Davis
Mauri	Valtonen	U. Turku
Tejaswi	Venumadhav	Institute for Advanced Study
Peter	Veres	University of Alabama in Huntsville
Francesco	Verrecchia	SSDC, INAF-OAR
Nevin	Weinberg	Massachusetts Institute of Technology
Sheng	Yang	oapd inaf/UC Davis
Michitoshi	Yoshida	Subaru Telescope, NAOJ
Qingjuan	Yu	Kavli Institute for Astronomy and Astrophysics
Adam	Zadrozny	University of Texas Rio Grande Valley

Gravitational Wave Astrophysics: Early Results from Gravitational Wave Searches and Electromagnetic Counterparts
Proceedings IAU Symposium No. 338, 2017
G. González & R. Hynes, eds.

doi:10.1017/S1743921318000169

Revealing Short GRB Jet Structure and Dynamics with Gravitational Wave Electromagnetic Counterparts

Gavin P. Lamb[1] **and Shiho Kobayashi**[1]

[1]Astrophysics Research Institute, Liverpool John Moores University, Liverpool, L3 5RF, UK
email: g.p.lamb@2010.ljmu.ac.uk

Abstract. Compact object mergers are promising candidates for the progenitor system of short gamma-ray bursts (GRBs). Using gravitational wave (GW) triggers to identify a merger, any electromagnetic (EM) counterparts from the jet can be used to constrain the dynamics and structure of short GRB jets. GW triggered searches could reveal a hidden population of optical transients associated with the short-lived jets from the merger object. If the population of merger-jets is dominated by low-Lorentz-factors, then a GW triggered search will reveal the on-axis orphan afterglows from these failed GRBs. By considering the EM counterparts from a jet, with or without the prompt GRB, the jet structure and dynamics can be constrained. By modelling the afterglow of various jet structures with viewing angle, we provide observable predictions for the on- and off- axis EM jet counterparts. The predictions provide an indication for the various features expected from the proposed jet structure models.

Keywords. gamma rays: bursts, gravitational waves

1. Introduction

Gravitational wave (GW) driven mergers of neutron star (NS) or black hole (BH) - neutron star binary systems are the most promising candidate for the progenitor of short duration γ-ray bursts (GRBs) (e.g. Eichler *et al.* 1989; Narayan *et al.* 1992; Mochkovitch *et al.* 1993; Bogomazov *et al.* 2007; Nakar 2007; Berger 2014). Mergering NS/BH-NS systems will be detectable by gravitational wave observatories where the mergers occur within the detection horizon (Nissanke *et al.* 2013). Amongst the electromagnetic (EM) counterparts are isotropic macro/kilo-nova (e.g. Li & Paczyński 1998; Barnes & Kasen 2013; Tanaka & Hotokezaka 2013; Piran *et al.* 2013; Metzger & Fernández 2014; Tanaka *et al.* 2014, 2017; Barnes *et al.* 2016; Tanaka 2016; Metzger 2017), radio counterparts (e.g. Nakar & Piran 2011; Kyutoku *et al.* 2014; Margalit & Piran 2015; Hotokezaka & Piran 2015; Hotokezaka *et al.* 2016), wide angle coccoon emission (Lazzati *et al.* 2017; Nakar & Piran 2017; Gottlieb *et al.* 2018; Kisaka *et al.* 2017), resonant shattering, merger-shock or precursor flares (Tsang *et al.* 2012; Kyutoku *et al.* 2014; Metzger & Zivancev 2016; Salafia *et al.* 2017), GRBs (e.g. Coward *et al.* 2014; Ghirlanda *et al.* 2016; Kathirgamaraju *et al.* 2017; Jin *et al.* 2017), failed GRBs (fGRB) (Dermer *et al.* 2000; Huang *et al.* 2002; Nakar & Piran 2003; Rhoads 2003; Lamb & Kobayashi 2016, 2017a,c), and off-axis orphan afterglows (e.g. Rossi *et al.* 2002; Granot *et al.* 2002; Zou *et al.* 2007; Zhang 2013; Lazzati *et al.* 2017; Sun *et al.* 2017).

GRBs are the dissipation of energy within relativistic jets launched due to the rapid accretion of material following the formation of a stellar mass BH (or a NS). Relativistic jets from accreting BH systems are seen on all mass scales, from the stellar mass BHs in some X-ray binaries and GRBs, to the super-massive BHs at the centres of galaxies. The universality of these jets indicates that the physical processes involved in the formation, collimation and acceleration are likely to be similar. These similarities have given rise to a number of attempts to show scaling relations for the observables from these systems (e.g Merloni *et al.* 2003; Heinz & Sunyaev 2003; Falcke *et al.* 2004; Türler *et al.* 2004; Yuan & Zhang 2012; Nemmen *et al.* 2012; Ma *et al.* 2014; Lamb *et al.* 2017b; Wang & Dai 2017). For the transient jets associated with GRBs, external shocks form as the jet decelerates leading to an afterglow (Sari *et al.* 1998).

The highly variable non-thermal emission of GRBs can only be explained if the jet or outflow is ultrarelativistic, with typical bulk Lorentz factors $\Gamma \sim 100$ (Mészáros 2002, etc.). As the jet decelerates it produces a broadband afterglow, this emission is beamed within an angle $1/\Gamma$. When $1/\Gamma > \theta_j$, the jets half-opening angle, a break in the afterglow lightcurve will be seen. For short GRBs, where the arrival time duration for 90% of the prompt emission is < 2s, the opening angle of the jet is poorly constrained. The range of opening angles from short GRB afterglow break time measurements indicates a wide range of jet half-opening angles, $2°$ at the narrowest and $\geqslant 25°$ for the widest (Fong *et al.* 2015). Energy distribution (or structure) within the jet is usually assumed to be homogeneous.

As jets are collimated, the probabilty that a system with a bi-polar outflow is inclined towards an observer is $\sim \theta_j^2/2$. For an isotropic distribution of systems, the on-axis probability is small. However, for a GW detected system where GWs are strongest in the polar directions, there is a GW Malmquist bias towards merging systems with a low-inclination (Kochanek & Piran 1993; Nissanke *et al.* 2010; Schutz 2011). Figure 1 shows the probability for a GW detected system with a given inclination, and the cumlative fraction of events inclined within an inclination angle. For jets with a wide half-opening angle, or where the EM counterparts from the jet are bright at inclinations $20 - 40°$ e.g. structured jets, then the potential to observe a jet origin EM counterpart following a GW detected NS/BH-NS merger is reasonable, $\sim 0.2 - 0.6$ (Lamb & Kobayashi 2017c).

In §2 we discuss the expected excess in a population of merger-jets where the dominant fraction has a low-Γ. In §3 we describe the lightcurves from the afterglow of various structured jets and give an indication of how some structure models can enhance the fraction of bright optical transients associated with GW detected NS/BH-NS mergers. §4 gives a summary of §2 and §3 plus additional comments and discussion regarding a population of low-Γ or structured jets.

2. Low-Γ Jets - failed GRBs

Where baryon loading of a relativistic outflow is efficient the bulk Lorentz-factor of the jet will be low i.e. $\Gamma << 40$ (e.g. Lei *et al.* 2013), where Γ is very low the GRB will be suppressed resulting in a failed-GRB and an afterglow-like transient. Failed GRBs, or dirty fireballs, were first discussed by Dermer *et al.* (2000); Huang *et al.* (2002); Nakar & Piran (2003) and Rhoads (2003) as potential sources of X-ray, optical and radio transients similar in appearance to a GRB afterglow but without the prompt high-energy trigger (e.g. Cenko *et al.* 2013). A population of low-Γ jets from mergers may go undetected, GW astronomy provides a trigger to search for such failed-GRB transients from merger-jets (Lamb & Kobayashi 2016).

Relativistic outflows become optically thin at the photospheric radius, $R_p \propto (E/\Gamma)^{1/2}$, and the minimum variability timescale for the prompt γ-ray emission constrains the radius from which these high-energy photons are emitted, $R_d \sim c\delta t\Gamma^2$. For a bright GRB the dissipation radius should be above the photospheric radius, $R_d > R_p$, and $\Gamma \gtrsim 80 E_{51}^{1/5} \delta t_{-1}^{-2/5}$ where $E_{51} = E/(10^{51}$ erg) and $\delta t_{-1} = \delta t/(0.1$ s). If $R_d << R_p$ then

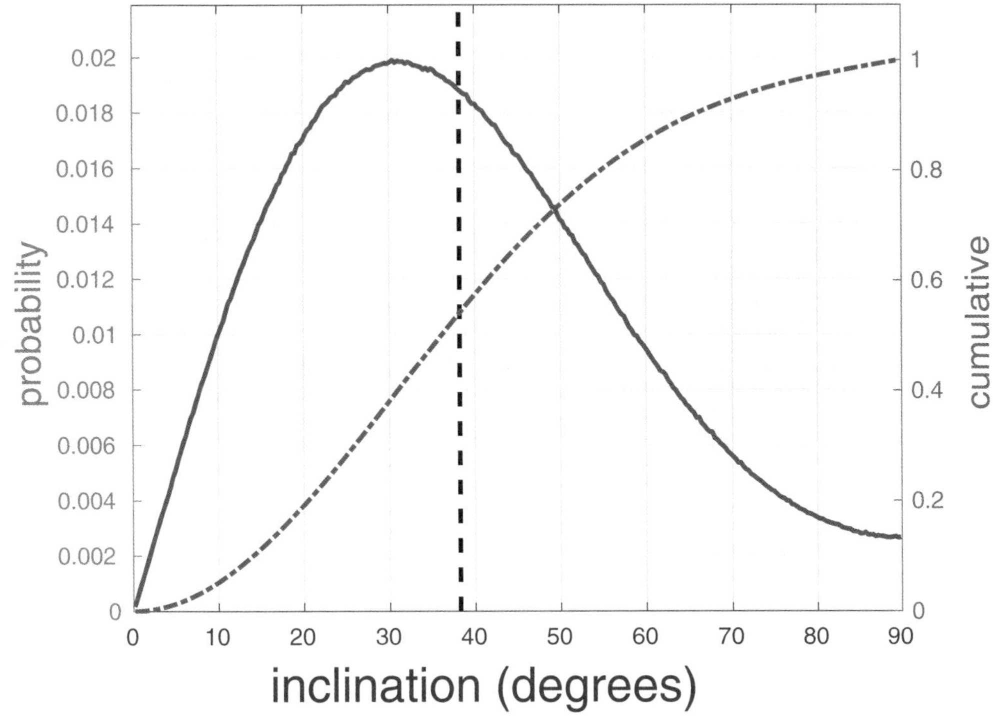

Figure 1. By considering the GW strain from a merger as a function of inclination, the distribution of system inclinations can be determined. For all GW detected mergers at a fraction of the maximum detectable luminosity distance, the probability of a system being inclined at given angle is shown with the blue solid line. The mean system inclination for this distribution is the dashed black line. The red dasheddotted line is the cumulative distribution. Figure from Lamb & Kobayashi 2017c

the γ-rays will be injected into an optically thick plasma. The high-energy photons will be coupled to the plasma and adiabatically cool until the optical depth reaches unity. Additionally the photons will undergo pair-production and Compton down-scattering that progressively thermalizes the distribution (Pe'er *et al.* 2005; Thompson *et al.* 2007; Hascoët *et al.* 2014).

To successfully produce a GRB at typical short GRB jet kinetic energies and efficiencies, the bulk Lorentz factor for an outflow should be $\Gamma \gtrsim 20-30$. For jets where the Lorentz-factor is below this limit, the outflow will not produce a GRB but will still result in a broadband afterglow that could be detected as an on-axis orphan afterglow. If we consider a cosmological population of merger jets that follow a Wanderman & Piran (2015) redshift and initial luminosity function, but where the bulk Lorentz-factor follows a distribution $N(\Gamma) \propto \Gamma^{-7/4}$, then $\sim 91\%$ of the population result in failed GRBs or GRBs too faint to be *Swift*/BAT detectable. Where the volume is limited to $z \leqslant 0.07$, approximately the face-on limit for NS-NS mergers detectable by advanced LIGO (Abbott *et al.* 2016, etc.) and there is an associated GW detection, then the fraction is $\sim 78\%$ failed GRBs (Lamb & Kobayashi 2016). These fractions only consider a population of systems with inclinations less than a jets half-opening angle.

The distribution of GRBs and failed GRBs with jet kinetic energy and bulk Lorentz-factor are shown in Figure 2 for the local events, $z \leqslant 0.07$. The inset shows an example afterglow lightcurve for a merger-jet at 40 Mpc, with a $\Gamma = 30$ and a jet kinetic energy

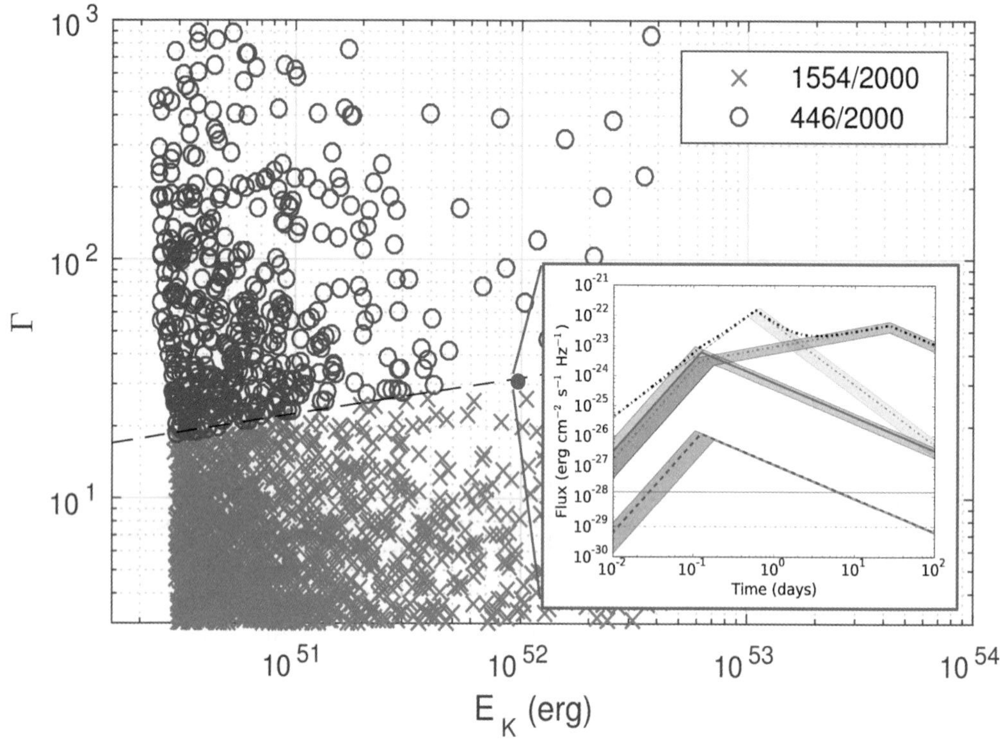

Figure 2. A Monte Carlo population of merger-jets oriented towards an observer within $z \leqslant 0.07$. Isotropic equivalent jet kinetic energy is shown on the x-axis, whilst the jet initial bulk Lorentz factor is shown on the y-axis. Blue circle indicates a *Swift*/BAT detectable short GRB, a red cross indicates a failed GRB. Inset: lightcurve for an on-axis low-luminosity GRB at 40 Mpc with parameters indicated by the grey circle. Red dashed-dotted line is forward- and reverse- shock emission at 10 GHz; green solid line is forward-shock r-band emission; and blue dashed line is forward-shock X-ray emission. Horizontal lines indicate flux level: 1μJy at 10 GHz, red dashed-dotted; and magnitude 21 at r-band, green solid line. The lower limit of the y-axis indicates the X-ray sensitivity of ~ 0.4 μCrab at 4 keV. Figure adapted from Lamb & Kobayashi 2016, 2017d

$E_{\rm k} \sim 10^{52}$ erg in an ambient medium with particle number density $n = (9 \pm 6) \times 10^{-3}$ cm^{-3}, the shaded regions represent the density uncertainty; such a merger-jet would produce a very low-luminosity GRB just below the detection limit for *Swift*/BAT. The afterglow at radio (10 GHz), optical (r-band) and X-ray (10^{18} Hz) are shown. The afterglow is bright for an on-axis observer (Lamb & Kobayashi 2017d).

For on-axis orphan afterglows at a distance less than ~ 300 Mpc, the optical peak flux is brighter than magnitude 21 for $\sim 85\%$ of cases whilst 10 GHz and X-ray are always brighter than the detection limits for the various facilities at these wavelengths (Lamb & Kobayashi 2016), e.g. VLA and XRT respectively. Optical and X-ray emission peaks on the same timescale, typically 0.1-10 days after a GW merger signal. Radio emission will peak from 10 days after the merger. Jets with a lower Lorentz-factor will peak later and be fainter due to the characteristic frequency being below that of the observation frequency at the peak time.

3. Jet Structure

Structured jets have a jet energy per steradian (or other parameter) that varies with angle from the central axis (e.g. Granot *et al.* 2002; Zhang & Mészáros 2002; Rossi *et al.* 2002, 2004; Vlahakis *et al.* 2003; Wei & Jin 2003; Kumar & Granot 2003; Panaitescu 2005; Peng *et al.* 2005; Jin *et al.* 2007; Barkov & Pozanenko 2011). A relativistic jet may have an intrinsic structure due to the formation mechanism (e.g. van Putten & Levinson 2003; Vlahakis *et al.* 2003) or as the jet breaks out of the medium immediately around the central-engine (e.g. Lyutikov & Blandford 2002; Levinson & Eichler 2003; Zhang *et al.* 2003, 2004; Lazzati & Begelman 2005; Morsony *et al.* 2010; Pescalli *et al.* 2015). As a jet breaks out from the dynamical ejecta associated with a NS/BH-NS merger, the jet will lose the collimating pressure of a cocoon (Bromberg *et al.* 2011). This may result in the wider components of the jet having a lower energy or Lorentz-factor distribution with angle from the jet-core region; such jet structures have been proposed as potential EM counterparts to GW detected NS/BH-NS mergers (e.g. Lamb & Kobayashi 2017c; Kathirgamaraju *et al.* 2017; Jin *et al.* 2017; Xiao *et al.* 2017).

An orphan afterglow population can reveal jet structure. The characteristic lightcurves for the optical (observed r-band) emission from four jet structures are shown in Figure 3. The lightcurves show afterglow at various inclinations for jets with a given structure: homogeneous jets, where the jet has a uniform energy and Lorentz-factor with angle from the central axis; two-component jets, where an energetic and fast core is surrounded by a wider sheath component with a fraction of the core energy; power-law jets, where the energy and Lorentz-factor reduce with angle from the core edge following a negative index power-law; and Gaussian jets, where the jet parameters follow a Gaussian distribution with angle from the central axis. A detailed description of these models is given in Lamb & Kobayashi (2017c).

Where wider jet components have a low-Lorentz factor, the prompt GRB emission is suppressed similar to a low-Γ jet case. In Figure 3 the lightcurves associated with an inclination that also produces a detectable GRB are shown as blue solid lines, failed-GRB afterglows as red dashed-dotted lines, and off-axis orphan afterglows as black dashed-dotted lines. The presence of jet structure is revealed at inclinations greater than the jet core angle. Thus orphan afterglows from NS/BH-NS mergers can be used to indicate the presence of extended jet structure beyond the homogeneous model.

For a GW detected population of NS/BH-NS mergers, the fraction of jet EM counterparts depends on the jet model. Where the jet population is made up of homogeneous jets with a half-opening angle $\theta_j = 6^\circ$, then the fraction of the population with a jet afterglow peak flux brighter than r-band magnitude 21 is $\sim 13.6\%$, of this fraction $\sim 13\%$ are GRB afterglows. Two-component jets have $\sim 30.0\%$ of the population brighter than magnitude 21, $\sim 9\%$ of these are GRB afterglows. Power-law jets $\sim 36.9\%$ of the population brighter than magnitude 21, where $\sim 59\%$ of these are GRB afterglows. Gaussian jets $\sim 13.3\%$ of the population brighter than magnitude 21, where $\sim 74\%$ of these are GRB afterglows. Here a GRB is defined as being detectable by *Swift*/BAT if the merger occurs within the instruments field-of-view.

4. Summary

EM counterparts from relativistic merger-jets accompanying GW detected NS/BH-NS mergers will reveal the structure and dynamical properties of short GRB jets. If a significant population of merger-jets result in collimated low-Γ outflows, then a hidden population of afterglow-like transients will be revealed. Such a population can be used

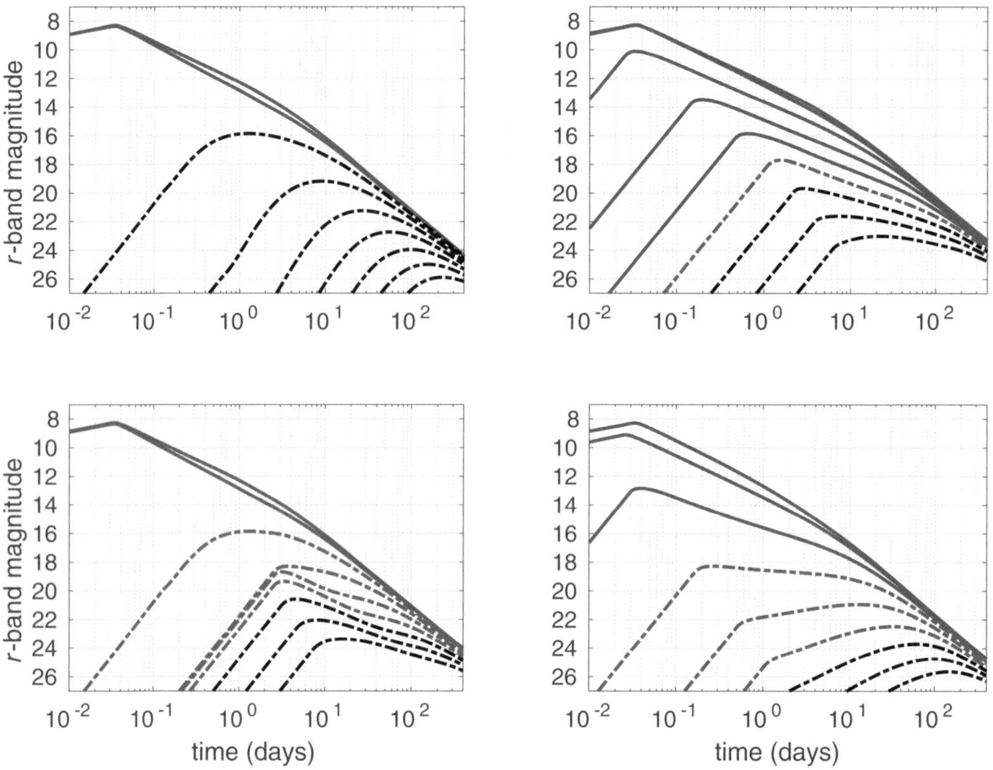

Figure 3. Afterglow r-band lightcurves for jets at 200 Mpc. Lightcurves are plotted for an observer at 5° increments in the range $0^\circ \leqslant \theta_{\rm obs} \leqslant 40^\circ$. The model values used in each case are: (top left) $\theta_c = \theta_j = 6^\circ$ for the homogeneous jet; (bottom left) $\theta_c = 6^\circ$ for the two-component jet where the second component extends to $\theta_j = 25^\circ$ with 5% of the core energy and Lorentz factor; (top right) $\theta_c = 6^\circ$ for the power-law jet with an index $k = 2$ for $\theta_c < \theta \leqslant 25^\circ$; and (bottom right) $\theta_c = 6^\circ$ for the Gaussian jet with a maximum $\theta_j = 25^\circ$. Jets have an isotropic equivalent blast energy of 2×10^{52} erg, a bulk Lorentz factor $\Gamma = 100$ for the core region, and an ambient medium density of $n = 0.1$. Blue lines indicate the afterglow of a GRB; red dashed lines indicate an on-axis orphan afterglow i.e. within the wider jet opening angle but with suppressed prompt emission; black dashed-dotted lines indicate an off-axis orphan afterglow. Figure adapted from Lamb & Kobayashi 2017c

to constrain the Lorentz-factor distribution of a population of merger-jets. Alternatively, the X-ray, optical or radio afterglow from a jet will reveal the presence of structure where the system is favourably inclined, i.e. $i \sim 20 - 40^\circ$. Sharp lightcurve peaks, re-brighening of the afterglow during the decline after peak, or a shallow rise index pre-peak are all signatures of a structured jet viewed at an inclination greater than the core angle.

For jets inclined at angles much greater than the γ-ray bright core region an associated GRB detection is not expected. However, the scattering of the prompt emission via a cocoon of a mergers dynamical ejecta could result in a faint GRB seen at such wide inclinations (Kisaka *et al.* 2017). Such a low-luminosity GRB will have an afterglow seen off-axis and peaking at ~ 100 days. The shape of the lightcurve for this afterglow will reveal any intrinsic jet structure.

A significant population of failed GRBs from merger-jets, or the presence of extended structure beyond a γ-ray bright jet-core, increases the rate of optical transients in an untriggered deep, $m \lesssim 26$, optical survey (e.g. LSST). These jet structures and dynamical qualities can equally be applied to a long GRB population. For a discussion of the transient rates from such jets for optical surveys see Lamb *et al.* (2017e).

Acknowledgements

GPL was able to attend the symposium due to International Astronomical Union (IAU) and Royal Astronomical Society (RAS) travel grants.

References

Abbott, B. P., Abbott, R., Abbott, T. D., *et al.* 2016, Living Reviews in Relativity, 19, 1
Barkov, M. V. & Pozanenko, A. S. 2011, MNRAS, 417, 2161
Barnes, J. & Kasen, D. 2013, ApJ, 775, 18
Barnes, J., Kasen, D., Wu, M.-R., & Martínez-Pinedo, G. 2016, ApJ, 829, 110
Berger, E. 2014, ARA&A, 52, 43
Bogomazov, A. I., Lipunov, V. M., & Tutukov, A. V. 2007, Astronomy Reports, 51, 308
Bromberg, O., Nakar, E., Piran, T., & Sari, R. 2011, ApJ, 740, 100
Cenko, S. B., Kulkarni, S. R., Horesh, A., *et al.* 2013, ApJ, 769, 130
Coward, D. M., Branchesi, M., Howell, E. J., Lasky, P. D., & Böer, M. 2014, MNRAS, 445, 3575
Dermer, C. D., Chiang, J., & Mitman, K. E. 2000, ApJ, 537, 785
Eichler, D., Livio, M., Piran, T., & Schramm, D. N. 1989, Nature, 340, 126
Falcke, H., Körding, E., & Markoff, S. 2004, A&A, 414, 895
Fong, W., Berger, E., Margutti, R., & Zauderer, B. A. 2015, ApJ, 815, 102
Ghirlanda, G., Salafia, O. S., Pescalli, A., *et al.* 2016, A&A, 594, A84
Gottlieb, O., Nakar, E., & Piran, T. 2018, MNRAS, 473, 576
Granot, J., Panaitescu, A., Kumar, P., & Woosley, S. E. 2002, ApJ, 570, L61
Hascoët, R., Beloborodov, A. M., Daigne, F., & Mochkovitch, R. 2014, ApJ, 782, 5
Heinz, S. & Sunyaev, R. A. 2003, MNRAS, 343, L59
Hotokezaka, K. & Piran, T. 2015, MNRAS, 450, 1430
Hotokezaka, K., Nissanke, S., Hallinan, G., *et al.* 2016, ApJ, 831, 190
Huang, Y. F., Dai, Z. G., & Lu, T. 2002, MNRAS, 332, 735
Jin, Z. P., Yan, T., Fan, Y. Z., & Wei, D. M. 2007, ApJ, 656, L57
Jin, Z.-P., Li, X., Wang, H., *et al.* 2017, arXiv:1708.07008
Kathirgamaraju, A., Barniol Duran, R., & Giannios, D. 2017, MNRAS, 473, 121
Kisaka, S., Ioka, K., Kashiyama, K., & Nakamura, T. 2017, arXiv:1711.00243
Kochanek, C. S. & Piran, T. 1993, ApJ, 417, L17
Kumar, P. & Granot, J. 2003, ApJ, 591, 1075
Kyutoku, K., Ioka, K., & Shibata, M. 2014, MNRAS, 437, L6
Lamb, G. P. & Kobayashi, S. 2016, ApJ, 829, 112
Lamb, G. P. & Kobayashi, S. 2017a, New Frontiers in Black Hole Astrophysics, 324, 66
Lamb, G. P., Kobayashi, S., & Pian, E. 2017b, MNRAS, 472, 475
Lamb, G. P. & Kobayashi, S. 2017c, MNRAS, 472, 4953
Lamb, G. P & Kobayashi, S. 2017d, arXiv:1710.05857
Lamb, G. P, Tanaka, M., & Kobayashi, S. 2017e, arXiv:1712.00418
Lazzati, D. & Begelman, M. C. 2005, ApJ, 629, 903
Lazzati, D., Deich, A., Morsony, B. J., & Workman, J. C. 2017, MNRAS, 471, 1652
Lei, W.-H., Zhang, B., & Liang, E.-W. 2013, ApJ, 765, 125
Levinson, A. & Eichler, D. 2003, ApJ, 594, L19
Li, L.-X. & Paczyński, B. 1998, ApJ, 507, L59
Lyutikov, M. & Blandford, R. 2002, Beaming and Jets in Gamma Ray Bursts, 146
Ma, R., Xie, F.-G., & Hou, S. 2014, ApJ, 780, L14
Margalit, B. & Piran, T. 2015, MNRAS, 452, 3419
Merloni, A., Heinz, S., & di Matteo, T. 2003, MNRAS, 345, 1057

Mészáros, P. 2002, ARA&A, 40, 137
Metzger, B. D. & Fernández, R. 2014, MNRAS, 441, 3444
Metzger, B. D. & Zivancev, C. 2016, MNRAS, 461, 4435
Metzger, B. D. 2017, Living Reviews in Relativity, 20, 3
Metzger, B. D. 2017, arXiv:1710.05931
Mochkovitch, R., Hernanz, M., Isern, J., & Martin, X. 1993, Nature, 361, 236
Morsony, B. J., Lazzati, D., & Begelman, M. C. 2010, ApJ, 723, 267
Nakar, E. & Piran, T. 2003, New Ast., 8, 141
Nakar, E. 2007, Phys. Rep., 442, 166
Nakar, E. & Piran, T. 2011, Nature, 478, 82
Nakar, E. & Piran, T. 2017, ApJ, 834, 28
Narayan, R., Paczynski, B., & Piran, T. 1992, ApJ, 395, L83
Nemmen, R. S., Georganopoulos, M., Guiriec, S., *et al.* 2012, Science, 338, 1445
Nissanke, S., Holz, D. E., Hughes, S. A., Dalal, N., & Sievers, J. L. 2010, ApJ, 725, 496
Nissanke, S., Kasliwal, M., & Georgieva, A. 2013, ApJ, 767, 124
Panaitescu, A. 2005, MNRAS, 363, 1409
Pe'er, A., Mészáros, P., & Rees, M. J. 2005, ApJ, 635, 476
Peng, F., Königl, A., & Granot, J. 2005, ApJ, 626, 966
Pescalli, A., Ghirlanda, G., Salafia, O. S., *et al.* 2015, MNRAS, 447, 1911
Piran, T., Nakar, E., & Rosswog, S. 2013, MNRAS, 430, 2121
Rhoads, J. E. 2003, ApJ, 591, 1097
Rossi, E., Lazzati, D., & Rees, M. J. 2002, MNRAS, 332, 945
Rossi, E. M., Lazzati, D., Salmonson, J. D., & Ghisellini, G. 2004, MNRAS, 354, 86
Salafia, O. S., Ghisellini, G., Ghirlanda, G., & Colpi, M. 2017, arXiv:1711.03112
Sari, R., Piran, T., & Narayan, R. 1998, ApJ, 497, L17
Schutz, B. F. 2011, Classical and Quantum Gravity, 28, 125023
Sun, H., Zhang, B., & Gao, H. 2017, ApJ, 835, 7
Tanaka, M. & Hotokezaka, K. 2013, ApJ, 775, 113
Tanaka, M., Hotokezaka, K., Kyutoku, K., *et al.* 2014, ApJ, 780, 31
Tanaka, M. 2016, Advances in Astronomy, 2016, 634197
Tanaka, M., Kato, D., Gaigalas, G., *et al.* 2017, arXiv:1708.09101
Tanaka, M., Utsumi, Y., Mazzali, P. A., *et al.* 2017, arXiv:1710.05850
Thompson, C., Mészáros, P., & Rees, M. J. 2007, ApJ, 666, 1012
Tsang, D., Read, J. S., Hinderer, T., Piro, A. L., & Bondarescu, R. 2012, Physical Review Letters, 108, 011102
Türler, M., Courvoisier, T. J.-L., Chaty, S., & Fuchs, Y. 2004, A&A, 415, L35
van Putten, M. H. P. M. & Levinson, A. 2003, ApJ, 584, 937
Vlahakis, N., Peng, F., & Königl, A. 2003, ApJ, 594, L23
Wanderman, D. & Piran, T. 2015, MNRAS, 448, 3026
Wang, F. Y. & Dai, Z. G. 2017, MNRAS, 470, 1101
Wei, D. M. & Jin, Z. P. 2003, A&A, 400, 415
Xiao, D., Liu, L.-D., Dai, Z.-G., & Wu, X.-F. 2017, arXiv:1710.00275
Yuan, F. & Zhang, B. 2012, ApJ, 757, 56
Zhang, B. & Mészáros, P. 2002, ApJ, 571, 876
Zhang, B. 2013, ApJ, 763, L22
Zhang, W., Woosley, S. E., & MacFadyen, A. I. 2003, ApJ, 586, 356
Zhang, W., Woosley, S. E., & Heger, A. 2004, ApJ, 608, 365
Zou, Y. C., Wu, X. F., & Dai, Z. G. 2007, A&A, 461, 115

Gravitational Wave Astrophysics: Early Results from Gravitational Wave Searches and Electromagnetic Counterparts
Proceedings IAU Symposium No. 338, 2017
G. González & R. Hynes, eds.

doi:10.1017/S1743921318000157

Gravitational wave optical counterpart searching based on GRAWITA and DLT40 project during LIGO O2 run

Sheng Yang[1,2]
on behalf of GRAWITA and DLT40 project.

[1]Department of Physics, University of California,
Davis, CA 95616-5270, USA
email: `sheng.yang@oapd.inaf.it`, `sngyang@ucdavis.edu`

[2]Osservatorio Astronomico di Padova,
INAF, I-35122 Padova, Italy

Abstract. The identification of the electromagnetic(EM) counterpart of gravitational wave(GW) trigger in the sky localization is a very difficult task because of the large uncertainty. Two complementary approaches are used in order to search for EM counterpart of GW signal with a typical large sky localization uncertainty: wide-field tilling search on high probability GW region, e.g. Gravitational Wave Inaf Team(GRAWITA) project or pointed search of selected galaxies in high probability GW region, e.g. Distance Less Than 40 Mpc survey(DLT40) project.

Keywords. gravitational wave: GW170817 — gamma-ray burst: GRB170817A — galaxies: NGC 4993 — kilonovae: TNS 2017 gfo, DLT17ck, SSS17a — multi-messenger astronomy

1. Introduction

GW, one of the fundamental predictions of general relativity, were finally detected on 14 September 2015 by the two Laser Interferometer gravitational-wave Observatory (LIGO) interferometers during the first LIGO run(O1)(Abbott *et al.* 2016a). Soon after, the second GW source was detected on 24 November 2015(Abbott *et al.* 2016b). GW is a very important and powerful tool to do astronomy, astrophysics and cosmology while to find EM counterpart of the GW events, which is so-called 'multi-messenger astronomy', will allow a precise localization of the GW source that will help streamline and enhance GW events by reducing the number of free parameters, or at least significantly restrict the range of their values that must be explored.

In the first, historic detection of GW150914, the two LIGO interferometers could achieve only a poor localization and hence EM counterpart searches had to scan a large area (10^3 square deg). With the beginning of joint LIGO and Virgo(Acernese *et al.* 2015) runs currently planned for the Fall 2016(Aasi *et al.* 2013), the simultaneous operations of 3 interferometers will secure a substantial reduction of the error areas. Nevertheless the error areas will still be dozens to hundred of square degrees and only cameras with very large field of view(FoV) can cover them entirely with a limited number of tiled pointings. One may wonder if it's possible to search the GW region with small telescopes whose answer is yes. Abadie *et al.*; Abadie *et al.*; Hanna *et al.* shows that searching the LIGO region based on galaxy catalog could greatly increases the chance of imaging an EM counterpart and the issue is to have a galaxy catalogue with a good completeness. GLADE† catalog, which is considered as the most complete galaxy catalogue so far, can

† **http://aquarius.elte.hu/glade/**

have a good completeness up to 70 Mpc, which means that we can use pointed strategy within this distance while once GW trigger happened further, the only choice would be the large FoV 'blind' search.

During the second run(O2) of the LIGO and Virgo Interferometer, a GW signal consistent with a binary neutron star coalescence (BNS) was detected on 2017 August 17th (GW170817) (Abbott *et al.* 2017a), quickly followed by a coincident short gamma-ray burst (GRB170817a(Fermi-GBM 2017)) trigger by the Fermi satellite (Meegan *et al.* 2009). About ten hours later, the DLT40 supernova search reported the discovery of a new optical transient, DLT17ck, as one of the six optical follow-up groups which independently detected this coincident optical kilonovae, 2017 gfo/sss17a (Abbott *et al.* 2017b; Coulter *et al.* 2017). DLT17ck was indeed cooling down and getting dimmer much faster than any other supernova(SN) we ever observed after monitoring the source for several days. Soon after, the spectra of this fast transient taken by some follow-up groups, including GRAWITA, proved that it's an r-process kilonovae which is characterized by rapidly expanding ejecta with spectral features similar to those predicted by current models. The era of multi-messenger astronomy has truly begun.

2. DLT40-like pointed galaxy search

Once the GW trigger occur in the local universe where the galaxy catalogue is still available, the galaxy pointing follow-up strategy, like DLT40, would be feasible.

DLT40 is a one day cadence supernova search in nearby galaxies aim to discover SN within 24 hours from explosion since this is the time when we can learn the most on the physics of the explosion. Within hundreds of square degrees, indeed, there is a limited number of nearby galaxies(Gehrels *et al.* 2016). The quick response of the Prompt telescopes (that we use for DLT40) is ideal to quickly observe the LIGO/Virgo region. A second advantage is that, since DLT40 is daily monitoring nearby galaxies†, we have useful information on those galaxies before detecting any variability. As soon as a new GW event is detected and the sky localization map is released. DLT40 would cross match the localization map with DLT40 galaxy catalog giving highest rank to bright galaxies coincident with high-probability map region. We then observed these galaxies for a period of 2 weeks for burst events and 3 weeks for merge events, which is related to the expected time evolution of optical counterpart of different LIGO events‡.

During the LIGO O2 period, on 2017 August 17 23:49:55 UT (11.09 hours after the LVC event GW170817), DLT40 detected DLT17ck, at RA=13:09:48.09 and DEC= -23:22:53.4.6, 5.37W, 8.60S arcsec offset from the center of NGC 4993 (Yang *et al.* 2017b, see Figure 1 Right Panel), as one of the six groups who indepandently detected this kilonova(Valenti *et al.* 2017). Yang *et al.* (2017a) used the observed light curve of DLT17ck(See Figure 3 Left panel) to constrain the rate of BNS mergers to less than $0.50\,^{+0.05}_{-0.04}$SNuB/$0.99\,^{+0.19}_{-0.15}$ 10^{-4} Mpc^{-3} yr^{-1} with SNe Ia extinction(See Figure 4 Left panel). After considering the completeness of the tracing mass and consuming time

† DLT40 is monitoring a sample of 2000 galaxies that have been selected among the galaxies of the Gravitational Wave Galaxy Catalogue (GWGC, White *et al.* (2011)). Starting from the GWGC catalogue, we selected only galaxies with recessional velocity $V \leqslant 3000km * s^{-1}$ (corresponding to $D \leqslant 40$ Mpc), with a declination ($Dec \leqslant +20$ deg), absolute magnitude ($MB \leqslant -18$ mag), and Milky Way extinction ($A_V \leqslant 0.5$ mag). Figure 1 Left panel shows all the 2000 DLT40 galaxy samples together with all SN found so far and the kilonoava, DLT17ck.

‡ burst events may be related to SN explosion that usually peak in two weeks. The merging of two neutron stars should produce an optical counterpart that has been predicted to be visible for few weeks (at least in red bands), but giving the high uncertainty of these models we select a conservative window of 3 weeks.

difference, we conclude that DLT40 need to be oprerated for $\sim$ 18.4years in order to independently discover a kilonova without GW information.

3. GRAWITA-like large FoV blind search

Once the GW trigger occur further where the galaxy catalogue completeness is poor, one may consider if the galaxy pointing follow-up strategy is still working and if not, the solution could be the large FoV blind follow-up strategy, like GRAWITA.

GRAWITA is an EM follow-up team by carrying out follow-up observational campaigns in the optical/NIR bands of the GW trigger observable with the VST, VLT, LBT, TNG, REM ground-based facilities. The VST 2.6m telescope(that GRAWITA used to follow the GW region) with 1 square degree FoV, can survey up to hundreds of square degrees every night and can reach to a limiting magnitude of r=22 mag on average by dithering 2 images with 45s exposure, which is very well designed for blind search following the GW area since it can reach deep enough and cover very large field(see Figure 2 Right panel). Also, most of the sources would be visible for GRAWITA at 410Mpc, which is the best value of GW150914 prediction, based on artificial star experiments towards different EM emission models(see Figure 4 Right panel). Figure 2 Left panel is an example by the VST 1 square degree survey boxes scheduled for GW150914. We performed the survey in r band with six epochs while the temporal sampling is chosen allowing for the rapid evolution of the expected GW's counterpart. Our immediate objective is to identify as early as possible all transients in the skymap area and select the best candidates for GW's counterparts. Once an interesting transient detected that can be link to a GW trigger, GRAWITA team and international collaborators would observe this source with an almost daily cadence in spectroscopic way, e.g. GRAWITA contributes to the spectroscopic classification of the kilonova, DLT17ck(Pian *et al.* 2017).

4. Implications and Future Prospects

Both strategies have played a very important role so far, in particular for the GW170817 event, no wonder that it's the success for the small FoV follow-up strategy but how about the future? Yang *et al.* (2017a) has such discussion that: During O3 run, LIGO and VIRGO will be able to detect BNS coalescences out to $65-85$ Mpc†. Considering all kilonovae detected later would be as bright as DLT17ck, while with the current DLT40 observing strategy, the limiting distance for DLT17ck is 70 Mpc. In order to cover the full LIGO/Virgo volume(in case one kilonova was detected in the limiting area), we would need to go $\sim$ 0.4 mag deeper, hence increasing the exposure time by a factor of 2.2, which means DLT40 could observe only $\sim$ 230 galaxies during a single night instead of current $\sim$ 500 galaxies. Even if we neglect that the GLADE catalog is only $\sim 85-90\%$ complete in the integrated luminosity up to 85 Mpc, the number of randomly selecting galaxies within 85 Mpc from the GLADE catalog in typical aLIGO/aVirgo regions (30 sq degrees) is $\sim$ 230, which is more or less the limiting number for DLT40. Therefore, within 85 Mpc, small telescopes can still play a useful role‡, but the incompleteness of the available catalogs, especially for faint galaxies, may suggest that a wide FoV tilling strategy may be preferred to avoid possible biases in sampling of the stellar population.

The wide FoV strategy would be the better and sometimes the only choice if the BNS source happened further more. From the GRAWITA wide FoV searching experiences, we

† We consider 85 Mpc in the discussion
‡ unless DLT17ck turns out to be a particularly bright kilonova

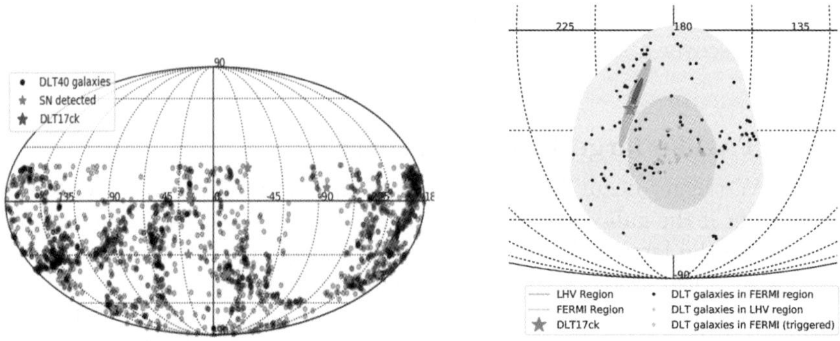

Figure 1. Left panel: DLT40 galaxy catalogue map with all SN detected by DLT40 so far and DLT17ck(Yang *et al.* 2017a). **Right panel**: The sky map region of GW170817 over-imposed on the Fermi GBM trigger GRB108017A. The red star marks the location of DLT17ck and the host galaxy NGC 4993(Valenti *et al.* 2017).

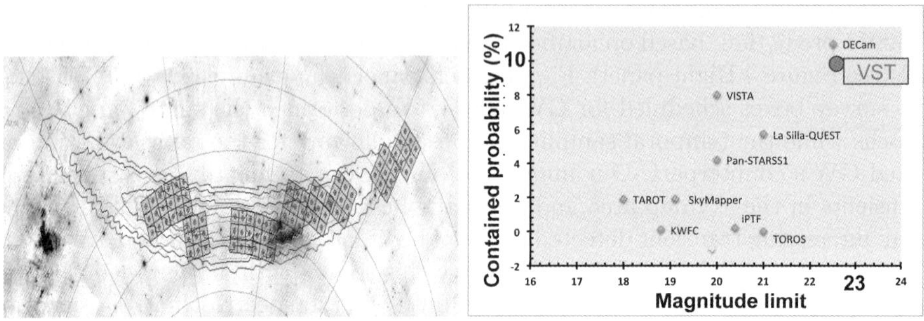

Figure 2. Left panel: VST 1 square degree survey boxes(90 pointings, 6 epochs) scheduled for GW150914.**Right panel**: VST performance: VST could include 10% probability of GW map with 22 mag in r band on average(Brocato *et al.* 2017).

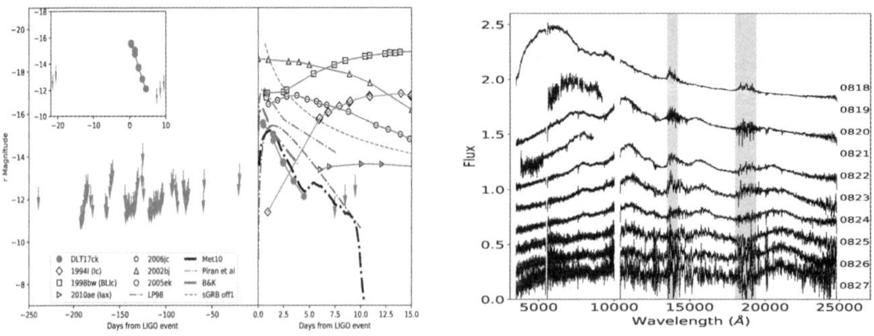

Figure 3. Left panel: DLT17ck light curve(in red) over plotted with normal or fast-evolving SNe (in gray). Several NS-NS merger models, scaled to a distance of 40 Mpc, are shown as comparison(Valenti *et al.* 2017). **Right panel**: The sequence of X-shooter, FORS2, and GMOS spectra for SSS17a/DLT17ck(Pian *et al.* 2017).

found that the difficulty of this strategy would be the large amount of consuming time for running pipelines and choose the best source as soon as possible during thousands of candidates. The parallel computing would increase the computing speed and instead of eyeballing candidates one by one, the machine learning algorithm would be the better choice and sometimes the only choice to rank the candidate list.

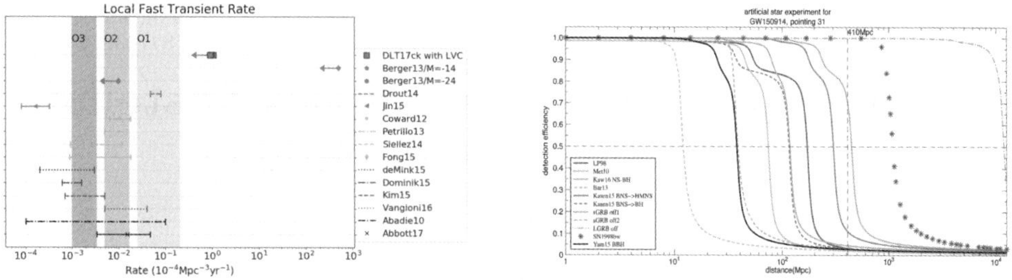

Figure 4. Left panel: DLT40 kilonova rate estimation, compared with the rate of sGRB(orange), BNS merger from stellar evolution(blue), cosmic nucleosyntesis(green), galactic pulsar population(magenta), gravitational waves(black) and fast optical transients(red)(Yang *et al.* 2017a). **Right panel**: How far VST can reach? With artificial star experiment, we test different kilonova models to estimate the limiting distance of VST images(Brocato *et al.* 2017).

With more interferometers joining in the future (e.g. KAGRA and LIGO-India), the GW sky localization would be more certain and the EM follow-up would become more effective, we believe that more multi-messenger sources would be detected and based on them would allow us to enrich our knowledge of the astrophysics in the multi-messenger astronomy era, thus, the past and the future in the precise cosmology erae.

References

Abadie, J., *et al.* 2012a, Astronomy & Astrophysics, 541, A155
Abadie, J., *et al.* 2012b, Astronomy & Astrophysics, 539, A124
Aasi, J., *et al.* 2013, Nature Photonics 7, 613
Aasi, J.,*et al.* 2015, Classical and Quantum Gravity, 32, 115012
Acernese, F., *et al.* 2015, Classical and Quantum Gravity, 32, 024001
Abbott, *et al.* 2016a, PRL, 116, 061102
Abbott, *et al.* 2016b, PRL, 116, 241103
Abbott, *et al.* 2017a, PRL, 119, 161101
Abbott, *et al.* 2017b, ApjL, 848, L12
Brocato, E., Branchesi, M., & Cappellaro, E. *et al.* 2017, MNRAS, arXiv:1710.05915
Coulter, D. A., Kilpatrick, C. D., & Siebert, M. R. *et al.* 2017, Sci, https://doi.org/10.1126/science.aap9811
Fermi-GBM 2017, GRB Coordinates Network, 524666471
Gehrels, N., Cannizzo, J. K. , & Kanner, J., 2016, Apj, 820, 136
Hanna, C., Mandel, I., & Vousden, W., 1993, Apj, 784, 8
Meegan, C., Lichti, G., & Bhat, P. N., 2009, ApJ, 702, 791
Pian, E., D'Avanzo, P., Benetti, S., *et al.* 2017, Nature, 551, 67
Tartaglia, L., Sand, D. J., Valenti, S., *et al.* 2017, arXiv:1711.03940
Valenti, S., David, J., Soung, S. *et al.*2017, apjl, 848, 2
White, D. J., Daw, E. J., & Dhillon, V. S. 2011, Classical and Quantum Gravity, 28, 085016
Yang, S., Valenti, S., & David, S. *et al.* 2017a, apjl, arXiv:1710.05864
Yang, S., Valenti, S., Sand, D., *et al.* 2017b, GRB Coordinates Network, 21531

Gravitational Wave Astrophysics: Early Results from Gravitational Wave Searches and Electromagnetic Counterparts
Proceedings IAU Symposium No. 338, 2017
G. González & R. Hynes, eds.

doi:10.1017/S1743921318003678

On the host galaxy properties of stellar binary black hole mergers

Youjun Lu*, **Liang Cao** and **Yuetong Zhao**

National Astronomical Observatories, Chinese Academy of Sciences, 20A Datun Road, Beijing 100012, China; * luyj@nao.cas.cn

School of Astronomy and Space Sciences, University of Chinese Academy of Sciences, 19A Yuquan Road, Beijing 100049, China

Abstract. We report our investigations on the host galaxy properties of stellar binary black holes (SBBHs) by implementing simple recipes for SBBH formation and merger into cosmological galaxy formation model. If the time delay between SBBH formation and merger ranges from Gyr to the Hubble time, SBBH mergers at redshift $z < 0.3$ occur preferentially in big galaxies with stellar mass $M_* > 2 \times 10^{10} M_\odot$ and metallicities Z peaking around $\sim 0.6 Z_\odot$. However, the host galaxy stellar mass distribution of heavy SBBH mergers (with total black hole mass $> 50 M_\odot$) is bimodal with one peak at $\sim 10^9 M_\odot$ and the other peak at $\sim 2 \times 10^{10} M_\odot$. The contribution fraction from metal-poor host galaxies ($Z < 0.2 Z_\odot$) to heavy mergers is much larger than that to less heavy mergers. If SBBHs were formed in the early universe, their mergers detected at $z < 0.3$ occur preferentially in even more massive galaxies with $M_* > 3 \times 10^{10} M_\odot$ and in galaxies with metallicities mostly $> 0.2 Z_\odot$ and peaking at $Z \sim 0.6 Z_\odot$.

Keywords. stars: black holes, gravitational waves, black hole physics.

1. Introduction

Gravitational wave (GW) events from the mergers of stellar binary black holes (SBBHs) are now expected to be regularly detected by the advanced LIGO (aLIGO) and VIRGO, examples include GW150914 [Abbott *et al.* (2016a)], GW151226 [Abbott *et al.* (2016b)], GW170104 [Abbott *et al.* (2017a)], GW170608 [Abbott *et al.* (2017b)], and GW170814 [Abbott *et al.* (2017c)]. Detections of those gravitational wave sources and many more in future, with precise measurements on physical parameters, offer a great tool to study the astrophysical origin of SBBHs and the abundant stellar and dynamical physics involved in, which are still not well understood, yet.

Several mechanisms have been proposed to produce SBBH GW sources. The leading mechanism is probably the evolution of massive binary stars in galactic fields in isolation [Mandel & de Mink(2016), Belczynski *et al.* (2016b)]. It is anticipated that the properties of the host galaxies of GW sources, if identified by future observations, can be used to reveal the formation mechanism for SBBHs and constrain the physics involved in the SBBH formation processes.

To identify the host galaxies of GW sources, one of the crucial ways is to find their electro-magnetic (EM) counterparts, if any, since the localization of these sources, obtained from GW signals only, is poor [Abbott *et al.* (2016c)]. Great efforts have been put into searching for EM counterparts of GW sources via broadband campaign (e.g., GW150914). However, there still seems little expectation of the detection of EM counterparts for SBBH mergers [Abbott *et al.* (2016d)]. If the host galaxy properties of GW sources can be known, the search for EM counterparts would be greatly narrowed. Therefore, it is of great importance to figure out where and when the GW sources were formed and what kind of galaxies they are hosted in at the GW detection time also by the means of searching for the EM counterparts of SBBH GW sources.

In this conference paper, we investigate the host galaxy properties of the SBBH GW sources by implementing simple SBBH formation recipes into a cosmological galaxy formation model. A simple SBBH formation model is introduced in Section 2. With this model, we show how to generate mock samples of SBBHs in section 3, with which the host galaxy properties are obtained and presented in section 4. Conclusions are given in section 5. Most of the results presented in this conference paper will be published in Cao *et al.* (2017).

2. Binary Black Holes formation and merger scenario

In general, the birth rate $R_{\rm birth}$ of single BHs with mass $m_\bullet$ per unit volume per unit time at the cosmic time t can be estimated as Abbott *et al.* (2016e), Dvorkin *et al.* (2016),

$$R_{\rm birth}(m_\bullet, t) = \int\int \dot{\psi}(Z;t)\phi(m_\star)\delta(m_\star - g_\bullet^{-1}(m_\bullet, Z))dm_\star dZ. \tag{2.1}$$

Here $\dot{\psi}(Z;t)$ is the star formation rate with metallicity Z per unit volume per unit time at the cosmic time t, $\phi(m_\star) \propto m_\star^{-\alpha}$ is the initial mass function (IMF) and the Chabrier IMF Chabrier(2003) is adopted below, δ is the Dirac-δ function, $m_\bullet = g_\bullet(m_\star, Z)$ is a function that describes the relation between the mass of a stellar remnant BH and the mass of its progenitor star and the latest version given in Spera *et al.* (2015) is adopted in this paper. The evolution time of a massive star to its remnant is usually on the order of $\leqslant 10^7$ yr and is ignored in the above equation. An SBBH may merger after a time period of $t_{\rm d}$ since its formation due to orbit decay by GW radiation. The GW event rate is then given by the convolution of the birthrate $R_{\rm birth}(m_{\bullet,1}, t)$ with the delay time distribution $P(t_{\rm d})$ and mass ratio distribution $P_q(q)$, i.e.,

$$R_{\rm GW}(m_{\bullet,1}, q; t) = f_{\rm eff} \int R_{\rm birth}(m_{\bullet,1}, t - t_{\rm d})P_q(q)P_t(t_{\rm d})dt_{\rm d}, \tag{2.2}$$

where $q = m_{\bullet,2}/m_{\bullet,1}$ is the ratio of the mass of the secondary black hole to that of the primary one, $f_{\rm eff}$ is the effective factor to form GW sources from binary stars.

With the above prescription, we can estimate the GW event rate of SBBH mergers by using the catalog of mock galaxies with detailed assembly histories given by Guo *et al.* (2011), including the star formation rate and the metallicity evolution of each mock galaxy. In this simple approach, all the physics governing the evolution of SBBHs are encoded in the three independent functions $f_{\rm eff}$, $P_q(q)$ and $P_t(t_{\rm d})$. Better estimates of the SBBH merger rate may be obtained by sophisticate binary evolution population synthesis model. However, there are still large uncertainties in the evolution models of massive (binary) stars, especially, the large uncertainties in the understanding of a number of physical processes, such as the common envelope evolution, the kick from supernova explosion, and the mass transfer, etc. [e.g., Dominik *et al.* (2013), Belczynski *et al.* (2016a), Mapelli *et al.* (2017)]. Different models may result in significantly different merger rate densities. Below we show that our simple approach can give similar merger rate density evolution compared against those obtained by using binary population synthesis codes.

Figure 1 shows the results on the merger rate density distribution as a function of redshift by adopting a reference model for $P_t(t_{\rm d})$, where $P(t_{\rm d} \propto 1/t_{\rm d}$ and $t_{\rm d} > 50\,{\rm Myr}$. For the calculation, $P(q)$ is set to $\propto q$ in the range from 0.5 to 1 according to binary population synthesis model results [e.g., Belczynski *et al.* (2016a)]; $f_{\rm eff}$ is set by calibrating the local merger rate to the observational constraint of $103{\rm Mpc}^{-1}{\rm yr}^{-1}$ [Abbott *et al.*

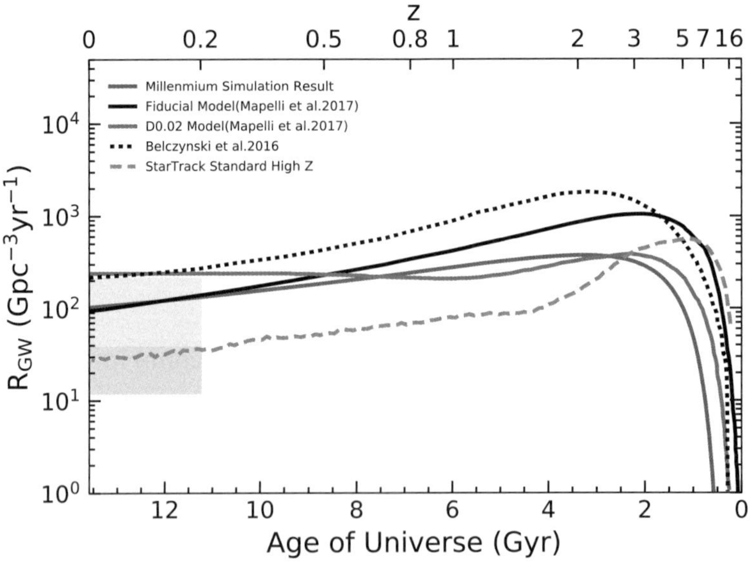

Figure 1. SBBH merger rate density as a function of the age of universe or redshift z. The red solid line represents the results obtained from the Millennium simulation galaxy catalog assuming the reference model for the time delay. The black dotted line represents the result obtained by Belczynski *et al.* (2016a) by using the extinction-corrected specific star formation rate in Madau & Dickinson (2014) and the binary population synthesis code StarTrack. The blue and green solid lines represent respectively the results from the fiducial model and the D0.02 model in Mapelli *et al.* (2017) by planting BBHs obtained from the binary population synthesis code SEVN into the Illustris simulations. The purple dashed line represents the merger rate density obtained by Dominik *et al.* (2013). The cyan and grey shaded regions represent the constraints on the merger rate density obtained from those detected GW sources by assuming two different IMF, respectively Abbott *et al.* (2017a).

(2017c)]. As seen from this figure, the merger rate density obtained by using the above simple recipes for SBBH formation and the New-Millennium cosmological galaxy catalog [Guo *et al.* (2011)] (red line) is more or less consistent with those obtained in previous works by Dominik *et al.* (2013), Belczynski *et al.* (2016a), and Belczynski *et al.* (2016a), by using binary population synthesis models. We note here that the large uncertainties in the evolution models of massive (binary) stars, especially, the large uncertainties in the understanding of a number of physical processes, such the common envelope evolution, the kick from supernova explosion, and the mass transfer etc. [e.g., Dominik *et al.* (2013), Belczynski *et al.* (2016a), Mapelli *et al.* (2017)] can lead to significantly different merger rate densities. The differences of the merger rate density estimated from our simple SBBH formation model from those estimates by using more sophisticated binary population synthesis models is comparable to the differences between those estimates in different works by using the population synthesis models.

3. Mock samples

Using the mock galaxy catalogs and the assembly and star formation history of each mock galaxy, we randomly assign SBBH merger GW events according to the probability for individual galaxies across cosmic time. We also impose that $\min[m_{\bullet,1}, m_{\bullet,2}] \geqslant 5M_{\odot}$, which is set by considering that all BHs measured dynamically have masses $> 5M_{\odot}$ [Özel *et al.* (2010), Farr *et al.* (2011)]. With these procedures, we can obtain mock catalogs

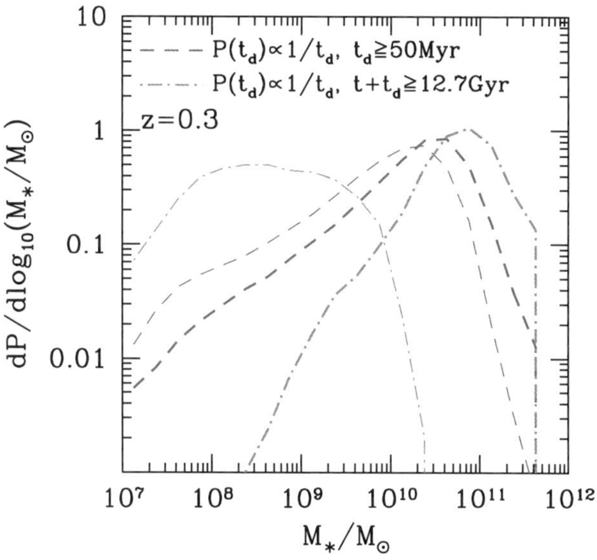

Figure 2. Stellar mass distribution of the host galaxies of binary black holes (SBBHs) at the SBBH formation time (thin lines) and merger time (thick lines). The red short-dashed and magenta dot-dashed lines show the results obtained from those time delay models $P(t_{\rm d}) \propto 1/t_{\rm d}$ with $t_{\rm d} \geqslant 50\,{\rm Myr}$, and $\geqslant 12.7\,{\rm Gyr} - t(z)$, i.e., the reference model, and the early SBBH formation model, respectively. Here $t(z) = \int_z^\infty \left|\frac{dt}{dz}\right| dz$ and $12.7\,{\rm Gyr}$ are the cosmic age at redshift z and at $z = 6$, respectively.

of SBBH GW events at any given cosmic time t (or correspondingly redshift z) for given $P_t(t_{\rm d})$ and $P_q(q)$. We generate mock catalogs of SBBH GW events at redshift $z = 0.3$, that enables the statistic studies on the properties of their host galaxies. Each GW event is characterized by the masses of the two components, i.e., $m_{\bullet,1}$ and $m_{\bullet,2} = qm_{\bullet,1}$, the merger time $t(z)$, the SBBH formation time $t(z) - t_{\rm d}$, the position and other properties (e.g., stellar mass M_*, metallicity Z_*) of its host galaxy, and the properties of its progenitor galaxy that the SBBHs formed in at the SBBH formation time $t - t_{\rm d}$.

4. Model Results

4.1. *Stellar mass distributions of SBBH host galaxies*

We extract the statistical information on their host galaxies from the mock catalogs for SBBH GW events and SBBHs. Figure 2 shows the stellar mass distribution of the host galaxies of those SBBHs at their merger time (i.e., the GW detection time $z = 0.3$) and formation time (thin lines), respectively. For the reference model, the distribution of the SBBH host galaxies at the GW detection time is shifted to higher masses compared with that at the SBBH formation time simply because of the growth of those host galaxies after the SBBH formation.

If the SBBHs were formed at early time, e.g., $z > 6$ (magenta dot-dashed lines for the early SBBH formation model), their host galaxies grew more significantly since their formation and became massive at the GW detection time.

4.2. *Host galaxies of SBBH mergers with different total masses*

SBBHs with different total masses $[M_{\bullet\bullet} = M_{\bullet,1} + M_{\bullet,2} = (1+q)M_{\bullet,1}]$ may have different formation histories. Heavy SBBHs may be formed only from metal poor binaries and

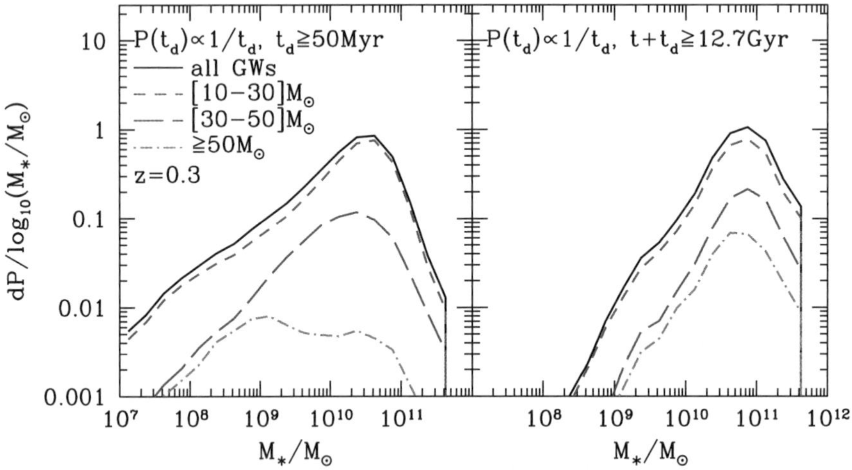

Figure 3. Stellar mass distributions of the host galaxies for SBBH mergers (at $z = 0.3$) with different total masses.

thus in metal poor galaxies at early times, therefore, they should be hosted in galaxies with different properties from those lighter SBBHs at the SBBH merger time.

Figure 3 shows the stellar mass distributions of the host galaxies of SBBH GW events with different $M_{\bullet\bullet}$ at $z = 0.3$. For the reference model, those heavy SBBHs with total mass $M_{\bullet\bullet} \geqslant 50M_{\odot}$, the host stellar mass distribution are broad and bimodal with one peak at $\sim 10^9 M_{\odot}$ and the other peak at $\sim 2 \times 10^{10} M_{\odot}$, which are corresponding to a population formed recently in small galaxies and another population formed at early time in small galaxies but merged into big galaxies later.

4.3. *Host galaxy metallicities*

Since the formation of BHs depends on the metallicity of their progenitor stars, the formation of SBBHs is also dependent on the metallicity of their progenitor binary stars and thus the metallicity of the progenitor galaxies when they formed in. However, the metallicity of the host galaxies at the GW detection time may be significantly different from that of the progenitor galaxies at the SBBH formation time.

Figure 4 shows the host galaxy metallicity distributions of those SBBHs at their merger time (thick lines) and at formation time (thin lines), respectively. The host galaxy metallicity distributions at the SBBH merger time shift to higher metallicities due to the metal enrichment after the SBBH formation.

If those SBBHs were formed at early time, e.g., $z > 6$ (the early SBBH formation model), then they were mostly formed in metal poor small galaxies. However, their host galaxies at the SBBH merger time ($z = 0.3$) have metallicities around $Z \sim 0.57 Z_{\odot}$.

Figure 5 shows the metallicity distributions of the host galaxies for SBBHs with different total mass ($M_{\bullet\bullet}$) range (i.e., $10 - 30M_{\odot}$, $30 - 50M_{\odot}$, and $\geqslant 50M_{\odot}$) at both the SBBH formation time and the SBBH merger time. For SBBH mergers with total mass $M_{\bullet\bullet} \geqslant 50M_{\odot}$, they must be formed in galaxies with metallicities $< 0.2 - 0.3Z_{\odot}$. However, the metallicities of their host galaxies at the SBBH merger time are substantially metal richer and have a distribution skewed towarding high metallicities if those SBBHs were formed at $z \geqslant 6$.

For light SBBHs, e.g., $M_{\bullet\bullet} = 10 - 30M_{\odot}$, they can be formed in galaxies with a large metallicity range. At the SBBH merger time, their host metallicities distribute over an even larger range with broader extensions at both the high Z and the low Z

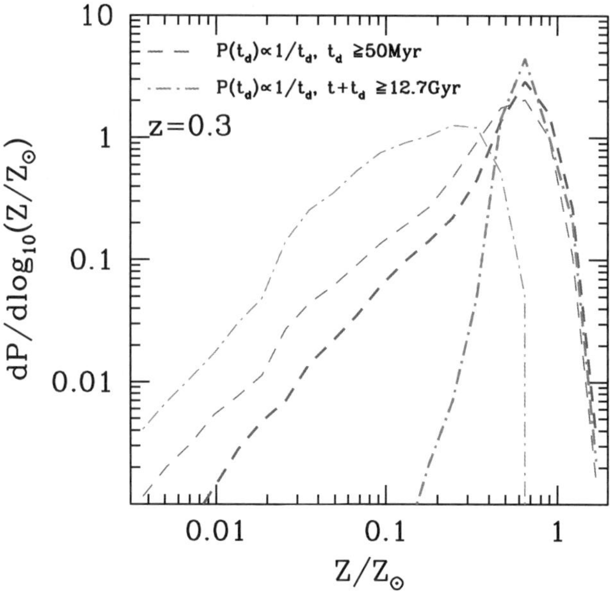

Figure 4. Metallicity distribution of the host galaxies at the SBBH formation time (thin lines) and the GW detection time or the SBBH merger time (thick lines). Legends for the lines and models are similar to Fig. 2.

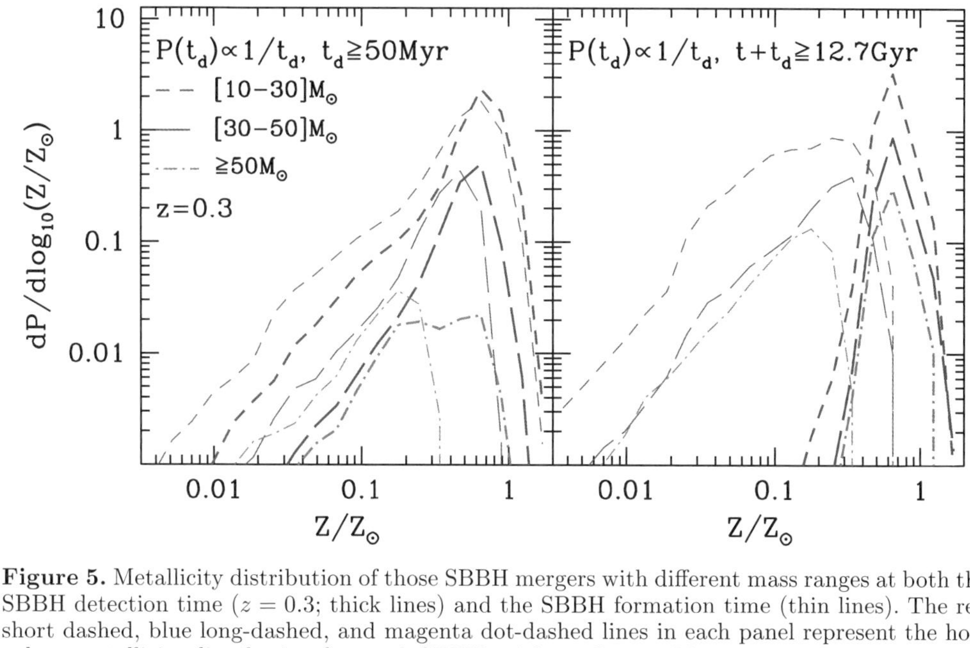

Figure 5. Metallicity distribution of those SBBH mergers with different mass ranges at both the SBBH detection time ($z = 0.3$; thick lines) and the SBBH formation time (thin lines). The red short dashed, blue long-dashed, and magenta dot-dashed lines in each panel represent the host galaxy metallicity distribution for mock SBBHs with total mass $M_{\bullet\bullet} = 10 - 30 M_{\odot}$, $30 - 50 M_{\odot}$, and $\geqslant 50 M_{\odot}$, respectively.

ends compared with that for $M_{\bullet\bullet} \geqslant 50 M_{\odot}$. For SBBHs with intermediate mass [$M_{\bullet\bullet} \in (30 - 50 M_{\odot})$], the distributions of the host metallicities at both the SBBH formation time and the SBBH merger time are in between those for the heavy SBBHs and for the light SBBHs (see Figure 5).

5. Conclusions

In this paper, we investigate the host galaxies of SBBH mergers, i.e., gravitational wave sources that may be detected by advanced LIGO and VIRGO, by implementing simple SBBH formation recipes into cosmological galaxy formation model using the Millennium-II simulation with a large box of side 137 Mpc, and we present a complete and thorough analysis of the properties of the SBBH host galaxy. We note here that understanding the host properties of SBBH mergers, as that done in this study and others Lamberts *et al.* (2016), O'Shaughnessy *et al.* (2017), Elbert *et al.* (2017), Schneider *et al.* (2017), is important for revealing the origin of those SBBHs and also idenfication of their EM counterparts. We summarize our main results as follows.

SBBH mergers with total mass $M_{\bullet\bullet} \geqslant 10M_{\odot}$ at low redshift ($z < 0.3$) occur preferentially in massive galaxies if the delay time between the SBBH formation and the merger is distributed in a broad range from less than Gyr to the Hubble time, and they occur preferentially in even more massive galaxies ($\sim 8 \times 10^{10} M_{\odot}$) if they were mostly formed at high redshift (e.g., $z > 6$) and have a large delay time (> 13 Gyr). For those heavy SBBH mergers ($M_{\bullet\bullet} \geqslant 50M_{\odot}$), their host stellar mass distribution is probably bimodal, with a low mass peak of $\sim 10^9 M_{\odot}$ and a high mass peak of $\sim 2 \times 10^{10} M_{\odot}$. The lower peak is mainly contributed by those SBBHs formed recently ($z < 0.5$), while the higher mass peak is mainly contributed by those SBBHs formed at early time ($z > 3.0$).

The formation of SBBHs depends on the metallicity of their progenitor binary stars. Heavy SBBHs ($M_{\bullet\bullet} \geqslant 50M_{\odot}$) were mostly formed in metal poor small galaxies with metallicities $< 0.2 - 0.3Z_{\odot}$. If the time delay between SBBH formation and merger covers a range from less than Gyr to the Hubble time, the host galaxy metallicity distribution of SBBH mergers detected at redshift $z < 0.3$ peaks around $0.6Z_{\odot}$ and has a long skewed wing towarding the low metallicity end ($Z < 0.2Z_{\odot}$). The host galaxy metallicity distribution of heavy SBBHs at the merger time at redshift $z < 0.3$ is ranging from $Z < 0.1Z_{\odot}$ to $\sim 1.0Z_{\odot}$, substantially broader compared with that for less heavy SBBHs. If SBBHs formed at early time, e.g., at $z > 6$, their mergers detected at $z < 0.3$ occur preferentially in galaxies mostly with metallicities $> 0.2Z_{\odot}$, because of significant metal enrichment of those host galaxies after the SBBH formation.

Acknowledgement

This work is partly supported by the National Natural Science Foundation of China under grant No. 11690024, 11373031 and 11390372, the Strategic Priority Program of the Chinese Academy of Sciences (Grant No. XDB 23040100), and the National Key Program for Science and Technology Research and Development (Grant No. 2016YFA0400704).

References

Abbott, B. P., Abbott, R., Abbott, T. D., *et al.* 2016, Physical Review Letters, 116, 061102
Abbott, B. P., Abbott, R., Abbott, T. D., *et al.* 2016, Physical Review Letters, 116, 241103
Abbott, B. P., Abbott, R., Abbott, T. D., *et al.* 2016, Living Reviews in Relativity, 19, 1
Abbott, B. P., Abbott, R., Abbott, T. D., *et al.* 2016, APJL, 826, L13
Abbott, B. P., Abbott, R., Abbott, T. D., *et al.* 2016, Physical Review Letters, 116, 131102
Abbott, B. P., Abbott, R., Abbott, T. D., *et al.* 2017, Physical Review Letters, 118, 221101
The LIGO Scientific Collaboration, the Virgo Collaboration, Abbott, B. P., *et al.* 2017, arXiv:1711.05578
Abbott, R., *et al.* 2017, arXiv:1709.09660
Belczynski, K., Repetto, S., Holz, D. E., *et al.* 2016a, APJ, 819, 108
Belczynski, K., Holz, D. E., Bulik, T., & O'Shaughnessy, R. 2016b, Nature, 534, 512

Cao, L., Lu, Y., & Zhao, Y. 2017, arXiv:1711.09190
Chabrier, G. 2003, PASP, 115, 763
Dominik, M., Belczynski, K., Fryer, C., *et al.* 2013, APJ, 779, 72
Dvorkin, I., Vangioni, E., Silk, J., Uzan, J.-P., & Olive, K. A. 2016, MNRAS, 461, 3877
Elbert, O. D., Bullock, J. S., & Kaplinghat, M. 2017, arXiv:1703.02551
Farr, W. M., Sravan, N., Cantrell, A., *et al.* 2011, APJ, 741, 103
Guo, Q., White, S., Boylan-Kolchin, M., *et al.* 2011, MNRAS, 413, 101
Lamberts, A., Garrison-Kimmel, S., Clausen, D. R., & Hopkins, P. F. 2016, MNRAS, 463, L31
Madau, P. & Dickinson, M. 2014, ARAA, 52, 415M
Mandel, I. & de Mink, S. E. 2016, MNRAS, 458, 2634
Mapelli, M., Giacobbo, N., Ripamonti, E., & Spera, M. 2017, arXiv:1708.05722
O'Shaughnessy, R., Bellovary, J. M., Brooks, A., *et al.* 2017, MNRAS, 464, 2831
Özel, F., Psaltis, D., Narayan, R., & McClintock, J. E. 2010, APJ, 725, 1918
Schneider, R., Graziani, L., Marassi, S., *et al.* 2017, arXiv:1705.06781
Spera, M., Mapelli, M., & Bressan, A. 2015, MNRAS, 451, 4086

Gravitational Wave Astrophysics: Early Results from Gravitational Wave Searches and Electromagnetic Counterparts
Proceedings IAU Symposium No. 338, 2017
G. González & R. Hynes, eds.

doi:10.1017/S1743921318003587

Reanalysis of LIGO black-hole coalescences with alternative prior assumptions

Davide Gerosa[1]†, Salvatore Vitale[2], Carl-Johan Haster[3], Katerina Chatziioannou[3] and Aaron Zimmerman[3]

[1]TAPIR 350-17, California Institute of Technology,
1200 E California Boulevard, Pasadena, CA 91125, USA
[2]LIGO, Massachusetts Institute of Technology,
77 Massachusetts Avenue, Cambridge, MA 02139, USA
[3]Canadian Institute for Theoretical Astrophysics, University of Toronto,
60 St. George Street, Toronto, ON M5S 3H8, Canada
email: dgerosa@caltech.edu

Abstract. We present a critical reanalysis of the black-hole binary coalescences detected during LIGO's first observing run under different Bayesian prior assumptions. We summarize the main findings of Vitale *et al.* (2017) and show additional marginalized posterior distributions for some of the binaries' intrinsic parameters.

These findings were presented at IAU Symposium 338, held on October 16-19, 2017 in Baton Rouge, LA, USA.

---

1. The incredible story of the first LIGO and Virgo events

The first two observing runs of the advanced gravitational wave (GW) detectors LIGO and Virgo (Aasi *et al.* 2015; Acernese *et al.* 2015) are an incredible story of scientific achievement (and possibly a pinch of luck).

The first observing run (O1) lasted from September 2015 to January 2016, during which the two LIGO detectors in Hanford, WA and Livingston, LA collected about ~ 51.5 days hours of coincident data. Three detections were made during this first data taking period, all from black hole (BH) binary systems (Abbott *et al.* 2016a). A few days *before* the official start of the scientific run, the strong event GW150914 from a BH binary with total mass $\sim 65 M_\odot$ hit the two LIGO detectors with a signal-to-noise ratio (SNR) of ~ 24 (Abbott *et al.* 2016c). This first landmark observation of GWs is one of the greatest achievements in modern science and crowned with success more than 50 years of experimental and theoretical effort in building and operating GW detectors. A few months later, on a Christmas day evening in the US, LIGO detected a second signal from a lighter BH binary of $\sim 24 M_\odot$, GW151226 (Abbott *et al.* 2016b). A weaker and less significant event, LVT151012, was also detected during O1. The probability that LVT151012 was not instrumental noise is 87% (Abbott *et al.* 2016a). It turns out this is not enough to qualify for the "GW" stamp, and the event was designated to be a LIGO/Virgo trigger.

The second observing run (O2) lasted from November 2016 to August 2017. The earliest two detections announced during O2 curiously mirrored the O1 results. First, GW17014 was detected from a BH binary of $\sim 50 M_\odot$ similar to that of GW150914 (Abbott *et al.* 2017a). Later on, GW170608 was detected from a BH binary of $\sim 19 M_\odot$ which resembles GW151226 (Abbott *et al.* 2017b). The real surprises came towards the end of the data taking period. GW170814 was the first event detected simultaneously by the two LIGO

† Einstein Fellow

detectors in the US and the Virgo interferometer in Europe (Abbott *et al.* 2017c). Adding the third detector to the network allowed for a drastic reduction of the sky location error box (from $\sim 1000\,\mathrm{deg}^2$ to $\sim 100\,\mathrm{deg}^2$), opening the possibility of rapid electromagnetic follow-up campaigns. In a remarkable twist of events, this possibility was realized only three days later, when GW170817 hit the LIGO detectors and was not confidently detected by Virgo (Abbott *et al.* 2017d). Virgo however, was taking data at the time, meaning the signal came from close to the blind spot of the interferometer. This allowed for an incredibly accurate sky localization within $\sim 27\,\mathrm{deg}^2$. The estimated masses of GW170817 are compatible with those of neutron stars, whose merger is expected to produce a variety of electromagnetic signatures. Indeed, the same event was seen in gamma rays just ~ 1.7 s after merger and, in just a few hours, several observatories identified an optical transient in NGC 4993, a lenticular galaxy at a mere distance of ~ 40 Mpc (Abbott *et al.* 2017e). This triggered an extensive follow-up campaign in all electromagnetic bands, providing us unique insights on neutron star mergers, short gamma-ray bursts, and kilonovae.

The parameters of all LIGO/Virgo events were estimated using powerful statistical pipelines which inevitably include prior assumptions. Here we present a critical reanalysis of the three O1 detections (GW150914, GW151226, and LVIT151012) under different Bayesian priors, summarizing results presented by Vitale *et al.* (2017). We repeat some references and context from Vitale *et al.* (2017) but refer the reader to that work for full details. This was the first independent reanalysis of the public LIGO data that made astrophysical inferences about the sources of the signals. This study pioneered the use of the scientific products released by the LIGO and Virgo Collaboration (losc.ligo.org, Vallisneri *et al.* 2015) to uncover finer details of these landmark discoveries.

2. Bayesian statistics: a magnifying lens for experimental data

Current BH binary data were analyzed using Bayesian statistics. In a nutshell, Bayesian statistics aims at improving our understanding of any given phenomenon (like BH mergers) by updating previous knowledge in light of new data. This principle is encoded in Bayes' theorem

$$P(\theta|d) = \frac{P(d|\theta)p(\theta)}{\int P(d|\theta)p(\theta)d\theta}\,. \tag{2.1}$$

The posterior probability $P(\theta|d)$ of measuring parameters θ in light of some new data d is proportional to both the likelihood $P(d|\theta)$ of measuring d and our prior beliefs $p(\theta)$. Much like a magnifying lens allows to spot finer details, Bayesian statistics is a natural procedure to make the information brought by the data evident against the background of our prior knowledge. Crucially, such background needs to be specified.

In GW research, we all carry a monumental prior, namely the theory of General Relativity (GR). GR, indeed, has passed all current experimental tests with flying colors. It is therefore reasonable to assume that GR is an accurate description of reality, even if it may not be the ultimate theory of gravity. This does not mean that GR is not put to the test with GW data, but rather that it is more reasonable to attempt measurements of *deviations from GR*, rather than measuring the absolute theory of gravity. This approach just reflects previous experience, corroborated by ~ 100 years of data, that saw GR coming out of any experimental test stronger than before.

3. What prior knowledge *could* go into a black hole analysis?

Astrophysical BHs (in GR at least) are fully characterized by two quantities, namely their mass and spin. A BH binary is therefore described by eight intrinsic parameters $\theta = \{m_1, m_2, \boldsymbol{\chi_1}, \boldsymbol{\chi_2}\}$. Let us describe the direction of each spin vector $\boldsymbol{\chi_i}$ with a polar angle θ_i measured from the binary's orbital angular momentum $\mathbf{L}$ and an azimuthal angle ϕ_i measured in the orbital plane. The relative orientations of the spins and the orbital angular momentum depend on the three angles θ_1, θ_2 and $\Delta\Phi = \phi_2 - \phi_1$.

Prior distributions have to be specified on all these parameters when analyzing BH coalescence data. All LIGO/Virgo analyses were performed with uniform prior distributions in $(m_1, m_2), \chi_1, \chi_2, \cos\theta_1, \cos\theta_2, \phi_1$ and ϕ_2 at the reference frequency of $f_{\rm ref} = 20$ Hz (Abbott *et al.* 2016a,b,c, 2017a,b,c,d). This is a very reasonable approach: we have never observed BHs in binaries before, so there is no reason to prefer a particular mass or spin range. However, this is surely not the only reasonable approach. For instance, spins are vectors, not scalar. One might prefer to assume spin vectors uniform in volume, rather than uniform in magnitude and isotropic in direction. We also have solid understanding that massive stars end their lives as BHs, following core collapse. Should we insert this piece of information into our BH analysis? The distribution of stellar masses follow a power-law distribution $p(m) \propto m^\alpha$ with $\alpha \sim -2.3$. Could this be a good candidate for a BH mass prior?

Vitale *et al.* (2017) first tried to address some of these issues by fully reanalyzing the O1 LIGO data with different prior assumptions. Their prior choices are summarized as follows:

(P_1) Masses are uniform, spin magnitudes are uniform, spin directions are isotropic. This is the standard prior choice made by the LIGO/Virgo Collaboration.

(P_2) Masses are uniform, spin magnitudes are uniform in rotational energy, spin directions are isotropic.

(P_3) Masses are uniform, spin magnitudes are uniform in volume, spin directions are isotropic.

(P_4) Masses are uniform, spin magnitudes are drawn from a bimodal distribution peaked at low and large spins, spin directions are isotropic.

(P_5) Masses are uniform, spin magnitudes are uniform, spin directions are preferentially aligned with $\mathbf{L}$ (i.e. small θ_i).

(P_6) Masses of the primary BHs are drawn from the stellar initial-mass function, secondary masses are uniform, spin magnitudes are uniform and spin directions are isotropic;

(P_7) Masses of the primary BHs are drawn from the stellar initial-mass function, mass ratio is drawn from a logistic distribution, spin magnitudes are uniform and spin directions are isotropic.

(P_8) Masses are uniform, spins magnitudes are preferentially low, spins directions are isotropic.

As detailed in Vitale *et al.* (2017), these choices leverage some know astrophysical properties of stellar-mass BHs and/or theoretical predictions on their population in binaries.

4. Marginalized posteriors under different priors

In Figs. 1 and 2 we shows marginalized posterior distributions for a variety of intrinsic parameters that characterize the observed BH events and all our prior choices P_1-P_8. Details on the data analysis technique are reported by Vitale *et al.* (2017). The two component masses m_1 and m_2 can be combined into total mass $M = m_1 + m_2$, mass

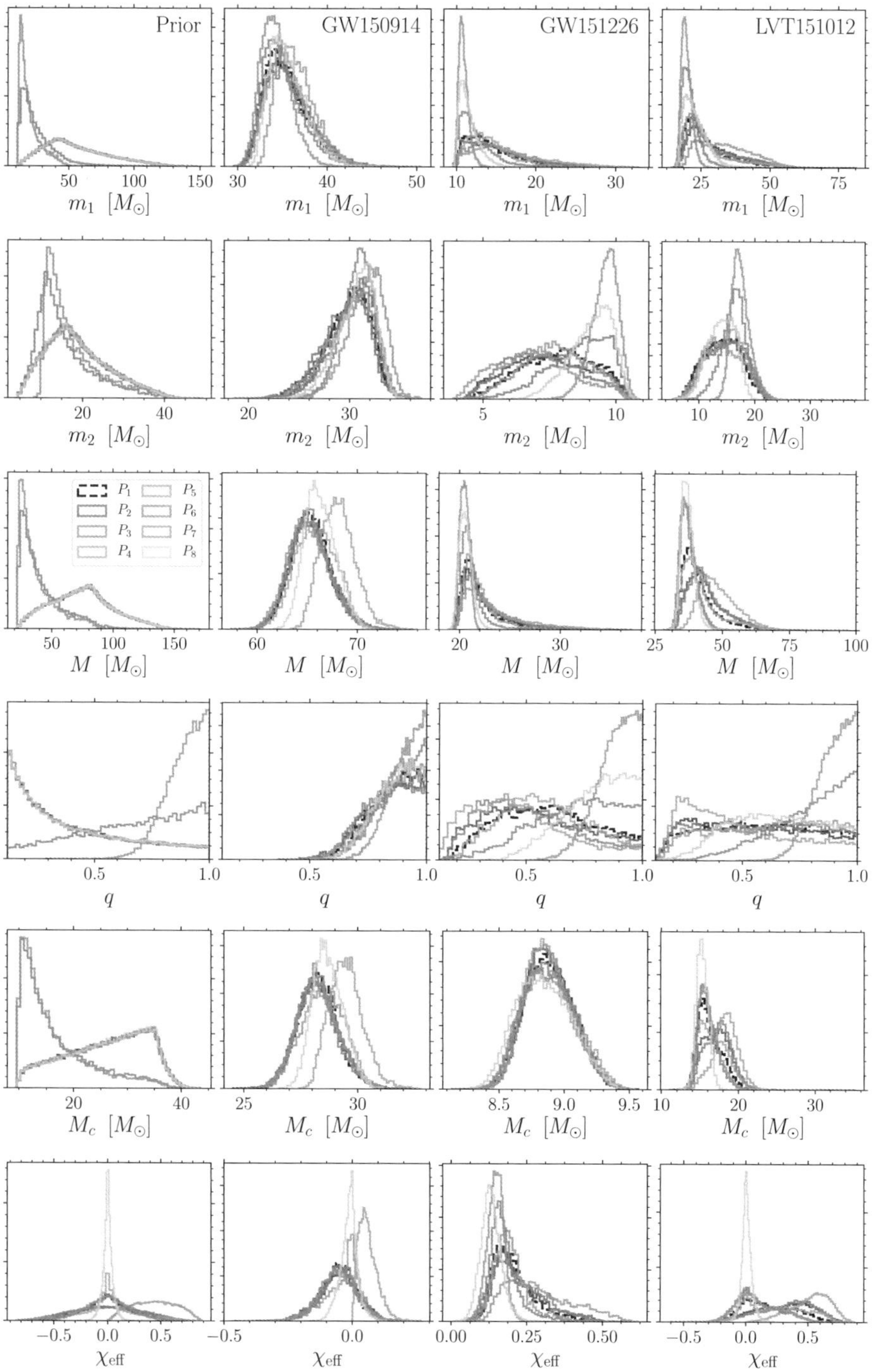

Figure 1. Prior (left panels) and posterior densities for GW150914, GW151226 and LVT151012 (second, third and fourth column from the left) under a variety of prior assumptions P_1-P_8. We show marginalized distributions for the BH component masses m_1 and m_2, total mass M, mass ratio q, chirp mass M_c and effective spin $\chi_{\rm eff}$. The $\chi_{\rm eff}$ panels were already presented by Vitale *et al.* (2017).

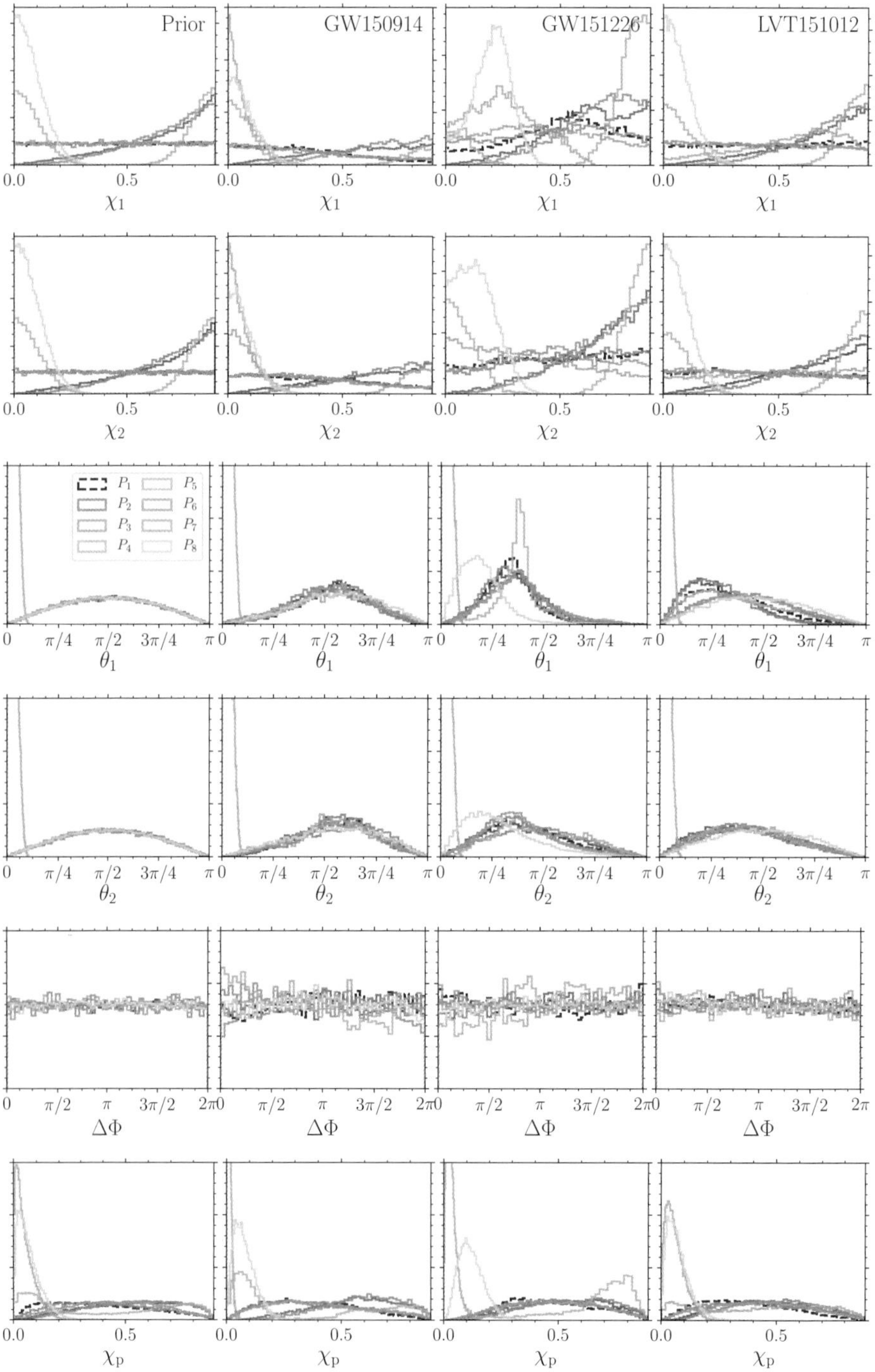

Figure 2. Prior (left panels) and posterior densities for GW150914, GW151226 and LVT151012 (second, third and fourth column from the left) under a variety of prior assumptions P_1-P_8. We show marginalized distributions for the BH spin magnitudes χ_1 and χ_2, the tilt angles θ_1 and θ_2 measured from the orbital angular momentum, the angle between the projections of the two spins onto the orbital plane $\Delta\Phi = \phi_2 - \phi_1$, and the precession parameter χ_p. Spin angles are reported at the reference frequency $f_{\rm ref} = 20$ Hz.

ratio $q = m_2/m_1 \leqslant 1$ and chirp mass $M_c = (m_1 m_2)^{3/5}/M^{1/5}$. The best measured spin quantity is the effective spin $\chi_{\rm eff} = (m_1\chi_1\cos\theta_1 + m_2\chi_2\cos\theta_2)/M$; spin precession is implemented into the waveform template used in this analysis through a single parameter χ_p.

Some of the inferred physical properties are robust under the choice of the Bayesian priors. For instance:

(a) non-spinning BHs for GW151226 are excluded at $> 90\%$ credible interval;
(b) conversely, GW150914 and LVT151012 can be described by $\chi_1 = \chi_2 = 0$;
(c) mass ratios $q \lesssim 0.5$ for GW150914 are excluded at $> 90\%$ credible interval;
(d) the 90% credible intervals in M_c change by $< 1 M_\odot$ in all cases;

On the other hand, other parameters strongly depend on the prior. Notably, this includes component masses m_i and spin magnitudes χ_i. The spin angles θ_1, θ_2, $\Delta\Phi$ and the spin precession parameter χ_p cannot be constrained meaningfully, under any of the priors tested here. We note, however, that this could be due to the waveform approximant used by Vitale *et al.* (2017) which only implements precession through a single effective spin. In general, prior effects are more severe for low SNR events like LVT151012, where data are less informative. A thorough discussion of these results (including odds ratio calculations) has been presented by Vitale *et al.* (2017). We encourage the reader to compare the findings spelled out by Vitale *et al.* (2017) with Figs. 1 and 2 of these proceedings.

5. Was it necessary?

Once a posterior $P(\theta|d)$ has been obtained from a reference prior $p(\theta)$ (say P_1 from above), one can in principle use Bayes theorem (2.1) to obtain the posterior $\tilde{P}(\theta|d)$ for a different prior $\tilde{p}(\theta)$:

$$\tilde{P}(\theta|d) \propto P(\theta|d)\frac{\tilde{p}(\theta)}{p(\theta)}\,. \tag{5.1}$$

This prompts the following question: was it necessary to fully reanalyze the data to obtain the posteriors presented in Sec. 4? Couldn't one just re-weight previous posterior samples as in Eq. (5.1)? While this procedure is mathematically correct, it might be dangerous in practice, because the numerical algorithm used to sample P might not cover the region of the parameter space that is of interest for $\tilde{P}$ densely enough. This issue was recently explored by Williamson *et al.* (2017): they find that re-weighted posterior samples might indeed lead to biassed conclusions. Re-weighting carries systematics which needs to be properly analyzed before Eq. (5.1) can be used in, e.g., hierarchical model selection schemes. Careful comparisons of our full reruns with re-weighted posteriors is a natural follow-up of the results presented by Vitale *et al.* (2017) and summarized here.

Acknowledgements

D.G. is supported by NASA through Einstein Postdoctoral Fellowship Grant No. PF6-170152 by the Chandra X-ray Center, operated by the Smithsonian Astrophysical Observatory for NASA under Contract NAS8-03060.

References

Aasi, J., *et al.* 2015, CQG, 32, 074001
Abbott, B. P., *et al.* 2016a, PRX, 6, 041015
—. 2016b, PRL, 116, 241103

—. 2016c, PRL, 116, 061102
—. 2017a, PRL, 118, 221101
—. 2017b, APJ, 851, L35
—. 2017c, PRL, 119, 141101
—. 2017d, PRL, 119, 161101
—. 2017e, ApJ, 848, L12
Acernese, F., *et al.* 2015, CQG, 32, 024001
Vallisneri, M., Kanner, J., Williams, R., Weinstein, A., & Stephens, B. 2015, in Journal of Physics Conference Series, Vol. 610, Journal of Physics Conference Series, 012021
Vitale, S., Gerosa, D., Haster, C.-J., Chatziioannou, K., & Zimmerman, A. 2017, PRL 119, 251103
Williamson, *et al.* 2017, PRD, 96, 124041

Gravitational Wave Astrophysics: Early Results from Gravitational Wave Searches and Electromagnetic Counterparts
Proceedings IAU Symposium No. 338, 2017
G. González & R. Hynes, eds.

doi:10.1017/S1743921318000170

High accuracy measurement of gravitational wave back-reaction in the OJ287 black hole binary

Mauri J. Valtonen[1,2], L. Dey[3], R. Hudec[4,5], S. Zola[6,7], A. Gopakumar[3], S. Mikkola[1], S. Ciprini[8,9], K. Matsumoto[10], K. Sadakane[10], M. Kidger[11], K. Gazeas[12], K. Nilsson[2], A. Berdyugin[1], V. Piirola[1,17], H. Jermak[13], K.S. Baliyan[14], D. E. Reichart[15], D. B. Caton[18], S. Haque[16] and the OJ287-15/16 Collaboration

[1]Department of Physics and Astronomy, University of Turku,
Väisäläntie 20, FIN-21500, Kaarina, Finland
email: mvaltonen2001@yahoo.com

[2]Finnish Centre for Astronomy with ESO, University of Turku,
Väisäläntie 20, FIN-21500 Kaarina, Finland

[3]Department of Astronomy and Astrophysics, Tata Institute of Fundamental Research,
Mumbai 400005, India

[4]Astronomical Institute, The Czech Academy of Sciences,
25165 Ondřejov, Czech Republic

[5]Czech Technical University in Prague, Faculty of Electrical Engineering,
Prague, Czech Republic

[6]Astronomical Observatory, Jagiellonian University,
ul. Orla 171, Cracow PL-30-244, Poland

[7]Mt. Suhora Astronomical Observatory, Pedagogical University,
ul. Podchorazych 2, PL30-084 Cracow, Poland

[8]Agenzia Spaziale Italiana (ASI) Space Science Data Center,
I-00133 Roma, Italy

[9]Istituto Nazionale di Fisica Nucleare, Sezione di Perugia,
I-06123 Perugia, Italy

[10]Astronomical Institute, Osaka Kyoiku University,
4-698 Asahigaoka, Kashiwara, Osaka 582-8582, Japan

[11]Herschel Science Centre, ESAC, European Space Agency,
28691 Villanueva de la Cañada, Madrid, Spain

[12]Department of Astrophysics, Astronomy and Mechanics, National & Kapodistrian University of Athens,
Zografos GR-15784, Athens, Greece

[13]Astrophysics Research Institute, Liverpool John Moores University,
IC2, Liverpool Science Park, Brownlow Hill, L3 5RF, UK

[14]Physical Research Laboratory,
Ahmedabad 380009, India

[15]University of North Carolina at Chapel Hill,
Chapel Hill, North Carolina NC 27599, USA

[16]Department of Physics, University of the West Indies,
St. Augustine, Trinidad&Tobago

[17]Kiepenheuer-Institut für Sonnenphysik,
D-79104 Freiburg, Germany

[18]Dark Sky Observatory, Department of Physics and Astronomy,
Appalachian State University, Boone, NC 28608, USA

Abstract. Blazar OJ287 exhibits large thermal flares at least twice every 12 years. The times of these flares have been predicted successfully using the model of a quasi-Keplerian eccentric black hole binary where the secondary impacts the accretion disk of the primary, creating the thermal flares. New measurements of the historical light curve have been combined with the observations of the 2015 November/December flare to identify the impact record since year 1886, and to constrain the orbit of the binary. The orbital solution shows that the binary period, now 12.062 years, is decreasing at the rate of 36 days per century. This corresponds to an energy loss to gravitational waves that is 6.5 ± 4 % less than the rate predicted by the standard quadrupolar gravitational wave (GW) emission. We show that the difference is due to higher order gravitational radiation reaction terms that include the dominant order tail contributions.

Keywords. black hole physics, gravitational waves, galaxies: BL Lacertae objects: individual (OJ287)

---

1. Introduction

OJ287 is a potential nHz gravitational wave source that could be observed by the Pulsar Timing Array method in near future when the detection sensitivity has increased by about a factor of three (Liu *et al.* 2012, Babak *et al.* 2016). The nucleus of this galaxy contains a binary black hole system with a 12 yr period and component masses 1.84×10^{10} $M_{\odot}$ and 1.5×10^{8} $M_{\odot}$. Its binary nature was discovered already in 1987 based on huge flares at about 12 yr intervals (Sillanpää *et al.* 1988, see Figure 1). Since then the arrival time of the flares has been predicted successfully in multiple occasions with the rms error of 16 days (See Table 1). In addition, it has been confirmed that the flares belong to the rare category of thermal flares (Valtonen *et al.* 2012, Valtonen *et al.* 2016). Therefore it is likely that the flares arise after a collision of the secondary black hole on the accretion disk of the primary (Ivanov *et al.* 1998, Pihajoki 2016) and it is these collisions that allow us to follow the orbital motion of the secondary around the primary black hole in great detail. The host galaxy of this system has V magnitude about 18 (Takalo *et al.* 1990, Nilsson 2017) which makes it similar to NGC 4889 in intrinsic brightness; the latter galaxy is one of the two bright central galaxies in the Coma cluster of galaxies, and is believed to harbor a supermassive black hole of mass similar to the primary in OJ287 (Graham and Scott 2013).

2. Overview

In recent years a great deal of new data have been added to the optical light curve of OJ287 (Hudec *et al.* 2013, Hudec 2017). They allow the recognition of 25 major flares since 1886 which may arise from disk impacts. Fourteen of them have good enough light curves to identify the start of the thermal flare (listed in Table 1) while two more are positioned on the time axis by dense observational upper limits (1906 and 1945 flares). Many have occurred during the Northern hemisphere summer period when OJ287 was not observed (especially in 1920's and 1930's) and cannot be used in this work. A set of nine flare timings is enough to solve the binary orbit. They determine uniquely the 8 parameters of our BBH central engine model: the two masses, the spin of the primary BH, the major axis, orbital eccentricity and the initial phase of the orbit, plus the two parameters of the standard accretion disk (Valtonen *et al.* 2010). With the addition of the 2015 flare timing we may explore an additional parameter which in this case is the magnitude of the gravitational wave tail term, not included in previous work. Once a solution for the orbit has been found using ten impact timings, we verify that the

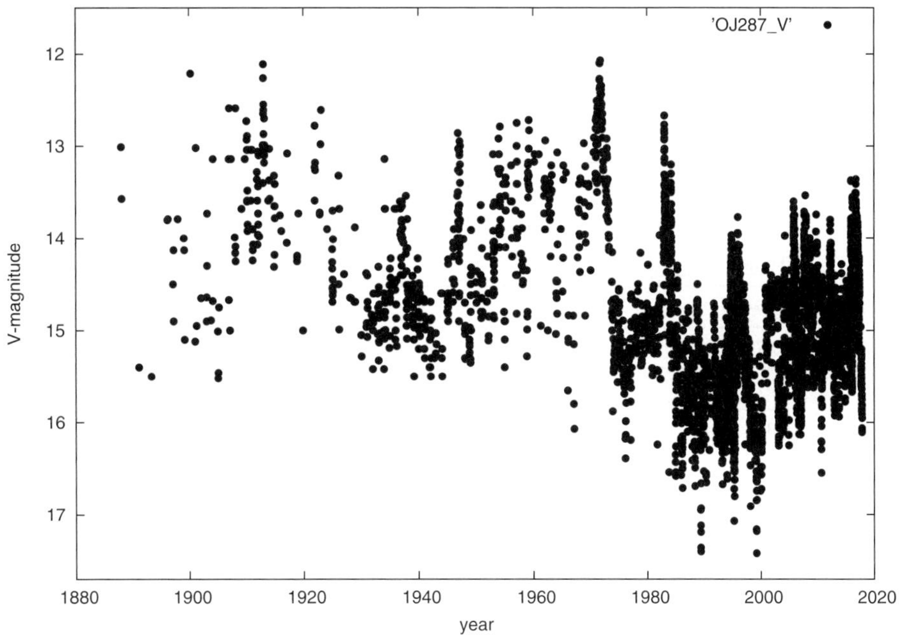

Figure 1. Historical light curve of OJ287. It has two dominant periodicities, 12 yr and 56 yr which imply the precession rate of $(12/56) \times 180$ degrees per period in the quasi-Keplerian orbit model. The corresponding mass of the primary is $\sim 1.8 \times 10^{10}\ M_{\odot}$. The light variations arise from the combination of varying Doppler boosting due to the wobble of the primary jet (Valtonen and Pihajoki 2013), from varying accretion rate due to tidal forces on the primary accretion disk (Sundelius *et al.* 1996, Valtonen *et al.* 2009), and from radiation arising from impacts of the secondary on the accretion disk (Lehto and Valtonen 1996).

solution is consistent with all observed flare times in Table 1, as well as the limits based on the observed upper limits. Thus we have altogether 16 constraints that have to be satisfied by a model with 9 parameters. Obviously, in general such problems do not have a solution, but if a solution is found, it is strongly overdetermined and produces very exact values for the parameters. In what follows, we briefly describe our PN-accurate description for BBH dynamics and how we determine the orbital parameters. Finally, we discuss the implications of our findings.

3. The Post-Newtonian orbit model

The relative acceleration between the two black holes in their center of mass frame, namely $\ddot{\boldsymbol{x}}$, can be divided in several PN contributions (Mora and Will 2004). Traditionally, the PN approximation provides the equations of motion of a binary as corrections to the Newtonian equations of motion in powers of $(v/c)^2 \sim GM/(c^2 R)$, where v, M, R and c are the characteristic orbital velocity, the total mass, the typical orbital separation of the binary, and speed of light, respectively. The relative acceleration and the precessional dynamics of the spin direction $\boldsymbol{s}_1$ are described by

$$\begin{aligned} \ddot{\boldsymbol{x}} \equiv \frac{d^2\boldsymbol{x}}{dt^2} &= \ddot{\boldsymbol{x}}_0 + \ddot{\boldsymbol{x}}_{1PN} + \ddot{\boldsymbol{x}}_{2PN} + \ddot{\boldsymbol{x}}_{2.5PN} \\ &\quad + \ddot{\boldsymbol{x}}_{SS} + \ddot{\boldsymbol{x}}_{3PN} + \ddot{\boldsymbol{x}}_{Q} + \ddot{\boldsymbol{x}}_{3.5PN} \\ &\quad + \ddot{\boldsymbol{x}}_{4PNtail} + \ddot{\boldsymbol{x}}_{SO} + \ddot{\boldsymbol{x}}_{4.5PN}\,, \end{aligned} \tag{3.1}$$

Table 1. Overview of quasi-Keplerian binary models of OJ287.

flare/ parameter	**1987-model** [1]	**1995-model** [2]	**2006-model** [3]	**2017-model** [4]
1913	1913.064	1912.124	1912.984	1912.981
1947	1948.014	1947.304	1947.264	1947.282
1957		1956.004	1957.104	1957.085
1959	1959.664	1959.234	1959.184	1959.212
1964		1963.794	1964.164	1964.226
1971	1971.314	1971.144	1971.104	1971.127
1973		1972.964	1972.954	1972.927
1983	1982.964 [5]	1982.964 [5]	1982.964 [5]	1982.964 [5]
1984		1984.124	1984.124	1984.119
1994	1994.614 [6]	1994.574	1994.604	1994.596 [7]
1995		1995.844 [6]	1995.824	1995.841 [8]
2005	2006.264 [6]	2005.704 [6]	2005.784	2005.744 [9]
2007		2007.724 [6]	2007.674 [6]	2007.691 [8]
2015	2017.914 [6]		2015.960 [6]	2015.875 [10]
primary $10^9 M_\odot$	18.4	17.7	18.25	18.35
secondary $10^9 M_\odot$	0.13	0.14 [11]	0.14 [12]	0.15
precession *deg*	38.6	33.3	39.1	38.7
spin χ	0	0	0 0.28 [12]	0.381

Notes:
[1] The 1987 tidal model was published in Sillanpää *et al.* (1988). The parameter values for this model are based on the two dominant frequencies of a quasi-Keplerian binary as determined by Valtonen *et al.* 2006 and matched with the tidal calculation by Sundelius *et al.* (1997). [2] The 1995 model was solved from timings of the secondary impacting on the accretion disk of the primary, and the time delay between the impact and an observed flare. The disk impact times come from Sundelius *et al.* (1997) and delay times from Lehto & Valtonen (1996; LV96), respectively. [3] The 2006 model was published in Valtonen (2007). The spin of the primary was assumed zero; a non-zero spin was added in the 2009 model (Valtonen *et al.* 2010). [4] The present 2017 model agrees completely with the observed flare times. [5] All models have been calibrated to the starting time of 1982.964 for the 1983 flare. [6] A prediction. The 2015 prediction uses the 2009 model. [7] The 1994 flare came within a week from the prediction, considering that the first (thermal) part of the flare was not observable due to closeness of OJ287 to the sun. [8] A note in the proofs of LV96 reports the observation of the predicted 1995 flare, exactly as expected. The 2007 flare came within days of the prediction (Valtonen *et al.* 2008). [9] The 2005 flare occurred two weeks late. [10] The 2015 flare was known to be spin sensitive. It occurred 4.5 weeks "too early", indicating an increase over the previous spin value. The rms deviation between observations and predictions has been 16 days in five cases. This contrast with the rms deviation for the predictions of the constant period 1987 model, 444 days. Any constant period model would give a similarly poor fit to observations. [11] Corrected for the Hubble constant H=70 km/s/Mpc. [12] For the 2009 model.

$$\frac{d\boldsymbol{s}_1}{dt} = (\boldsymbol{\Omega}_{SO} + \boldsymbol{\Omega}_{SS} + \boldsymbol{\Omega}_Q) \times \boldsymbol{s}_1 \,, \tag{3.2}$$

where we let $\boldsymbol{x} = \boldsymbol{x}_1 - \boldsymbol{x}_2$ as the center-of-mass relative separation vector between the black holes with masses M_1 and M_2. $\ddot{\boldsymbol{x}}_0$ represents the Newtonian acceleration given by $\ddot{\boldsymbol{x}}_0 = -\frac{G\,m}{r^3}\,\boldsymbol{x}$ where $m = M_1 + M_2$, $r = |\boldsymbol{x}|$. Also we define $\boldsymbol{n} \equiv \boldsymbol{x}/r$, $\dot{\boldsymbol{x}} = \boldsymbol{v}$ and $\eta = M_1\,M_2/m^2$. We include contributions due to general relativistic spin-orbt, spin-spin and classical spin orbit interactions, denoted by $\boldsymbol{\Omega}_{SO}$, $\boldsymbol{\Omega}_{SS}$ and $\boldsymbol{\Omega}_Q$, respectively in the precessional angular velocity for the primary BH spin.

The PN contributions occurring at the conservative 1PN, 2PN, 3PN and the reactive 2.5PN, 3.5PN and 4.5PN orders, denoted by $\ddot{\boldsymbol{x}}_{1PN}$, $\ddot{\boldsymbol{x}}_{2PN}$, $\ddot{\boldsymbol{x}}_{3PN}$, $\ddot{\boldsymbol{x}}_{2.5PN}$, $\ddot{\boldsymbol{x}}_{3.5PN}$ and $\ddot{\boldsymbol{x}}_{4.5PN}$, respectively, are non-spin by nature. The explicit expressions for these contributions suitable for describing the binary black hole dynamics in the modified harmonic gauge are found in Will and Maitra (2017).

The expression for the radiation reaction terms may be written as

$$\begin{aligned}
\ddot{\boldsymbol{x}}_{2.5PN} &= \frac{8}{5}\frac{G^2 m^2 \eta}{c^5 r^3}\left\{A_{2.5}\dot{r}\boldsymbol{n} - B_{2.5}\boldsymbol{v}\right\}\\
\ddot{\boldsymbol{x}}_{3.5PN} &= -\frac{8}{5}\frac{G^2 m^2 \eta}{c^7 r^3}\left\{A_{3.5}\dot{r}\boldsymbol{n} - B_{3.5}\boldsymbol{v}\right\}\\
\ddot{\boldsymbol{x}}_{4.5PN} &= \frac{8}{5}\frac{G^2 m^2 \eta}{c^5 r^3}\left\{A_{4.5}\dot{r}\boldsymbol{n} - B_{4.5}\boldsymbol{v}\right\}
\end{aligned}$$

where, for example, $A_{2.5} = 3v^2 + \frac{17}{3}\frac{G\,m}{r}$ and $B_{2.5} = v^2 + 3\frac{G\,m}{r}$ while the corresponding expressions for the 3.5 and 4.5 PN orders are more complicated (see Will and Maitra 2017). We find that when integrated over a quasi-Keplerian orbit the contribution from the A-coefficients are nearly symmetric but opposite in sign with respect to the pericenter. Therefore, they do not contribute much to the orbital averaged quantities. On the other hand, the B-coefficients, even though also symmetric with respect to the pericenter, make contributions of the same sign, and are thus important to us. The dominant order hereditary tail contributions to the reactive orbital dynamics are introduced by assuming that they are directly proportional to the "Newtonian" radiation term:

$$\ddot{\boldsymbol{x}}_{4PNtails} = radfac\;\ddot{\boldsymbol{x}}_{2.5PN}$$

This is mainly due to the absence for closed form expressions for the tail contributions to energy and angular momentum fluxes of an eccentric binary (Gopakumar *et al.* 1997). The value of $radfac$ is unknown and it is determined as one of the parameters of the orbit solution and there are on-going investigations to constrain it from theoretical considerations for our BH binary.

The leading order spin-orbit contributions to $\ddot{\boldsymbol{x}}$ appear at the 1.5PN order (Barker & O'Connell 1979) while the next higher order is at the 2.5 PN level (Will & Maitra 2017). These terms are included in $\ddot{\boldsymbol{x}}_{SO}$, and they are linear in the Kerr parameter χ. The Kerr parameter and the unit vector $\boldsymbol{s}_1$ define the spin of the primary black hole by the relation $\boldsymbol{S}_1 = G\,m_1^2\,\chi\,\boldsymbol{s}_1/c$. The Kerr parameter χ is allowed to take values between 0 and 1 in GR. The terms $\ddot{\boldsymbol{x}}_{SS}$ and $\ddot{\boldsymbol{x}}_Q$ include the leading order general relativistic spin-spin interactions and certain classical spin-orbit interactions that involve the quadrupole moment of the primary BH and they appear at the 3PN order for slowly rotating BHs (Will & Maitra 2017). The equations of motion includes certain reactive 4PN spin-orbit contribution, while the $\ddot{\boldsymbol{x}}_{4PNtail}$ term models the above mentioned leading order gravitational wave tail contributions.

Table 2. Parameters of the orbit solution.

Parameter		unit	error
M_1	18348	$10^6 M_\odot$	±101
M_2	150.13	$10^6 M_\odot$	±0.43
χ_1	0.381		±0.0025
h	0.845		±0.002
t_d	0.7734		±0.0015
$\Delta\Phi$	38.726	*deg*	±0.012
Θ_0	55.57	*deg*	±0.21
e_0	0.657		±0.001
P_{2017}	12.062	yr	±0.001
ΔP	36.2	d/100 yr	±0.25
$radfac$	1.304		±0.01
$\Delta 2.5PN$	−0.065		±0.04
α_g	0.1		±0.05

There are 12 degrees of freedom in the $\ddot{\boldsymbol{x}}_{4.5PN}$ term (Gopakumar *et al.* 1997). However, these arbitrary parameters completely drop out of the orbit averaged equations for the orbital elements (Will & Maitra 2017). This allows us to make a free choice of these parameters without significantly affecting the derived orbit, a fact that we have verified by experimenting with different combinations of those parameters. The spin of the primary black hole precesses due to the leading order general relativistic spin-orbit coupling as described by equation (3.2). The precessional equation for the unit spin vector $\boldsymbol{s}_1$ enters the binary dynamics essentially at the 2PN order.

4. Solving the orbit

The orbit search algorithm starts from a trial orbit, and if the adopted outburst times do not fit, it automatically adjusts itself until a good model orbit is found. There are no solutions in general, but if one is found, it is uniquely defined by the nine parameters of the model. The details of the algorithm are given in Valtonen (2007).

We find that a solution exists if only the first "Newtonian" radiation reaction term (PN2.5) is used. Adding the next level correction (at 3.5PN) leads to a loss of solution, and it is not helped by adding the 4.5PN level. Therefore it is necessary to incorporate the effects of 4PN tail contributions. Unfortunately, there exists no closed form expression for such tail contributions to the orbital dynamics of an eccentric binary. This is why we invoke an ambiguity parameter $radfac$ as a coefficient of our additional "Newtonian" radiation reaction term and use it as a free parameter in the orbit solution. The fifth column in Table 1 describes the best model with the current data. Most important of the nine independent parameters of the model are the two masses M_1 and M_2, the spin of the primary χ, and the precession rate of the major axis per period $\Delta\Phi$. Further, we also determine the eccentricity of the orbit e_0, and its orientation at a given epoch Θ_0. The two parameters of the standard accretion disk in the calculation are the disk thickness h and the delay time between an impact and a flare t_d, both given in units of Lehto and Valtonen (1996). It is possible to translate them to the more familiar disk parameters α_g(viscosity) and $\dot{m}$ (mass accretion rate). In addition, we determine the ambiguity parameter related to the 4PN term $radfac$. The other properties of interest are the present (redshifted) period of the orbit P_{2017} and the rate of decrease of the period ΔP. From the ambiguity parameter $radfac$ we also derive the correction to the "Newtonian" gravitational wave radiation reaction arising from higher order radiation

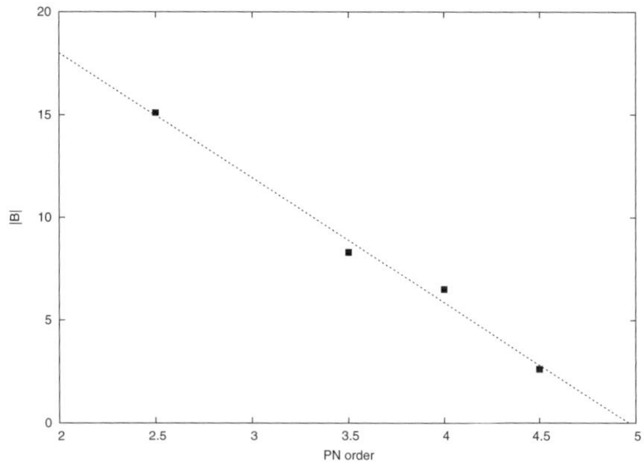

Figure 2. The absolute value of the B component of gravitational radiation term at the pericenter of the binary orbit, and at different Post Newtonian orders. The sign of the 3.5PN term is opposite of the other terms. Therfore the terms above 2.5PN nearly cancel each other at the pericenter.

terms $\Delta 2.5PN$. The mean values of these parameters and their standard deviations for 97 orbital solutions (found from 1000 trials) are listed in Table 2.

In Figure 2 the absolute values of the radiation reaction terms and the tail term are compared with each other at the pericenter of the binary orbit. We note that there is a monotonic decrease of the terms with increasing PN order. A linear regression suggests that GW emission induced 5PN contributions are negligible in the OJ287 problem, and thus we have the required level of accuracy in use.

5. Implications

A strongly over-determined orbital solution implies that its parameters are known very accurately, at better than one percent level for the masses and the primary BH spin. The orbit solution shows that the period of the orbit, now 12.062 year, has decreased at the rate of 36 days per century. This corresponds to an energy loss to gravitational waves that is 6.5 ± 4 % less than the rate predicted by the standard quadrupolar GW emission. We have shown that the difference is due to higher order gravitational radiation reaction contributions to BBH dynamics that includes the dominant order tail contributions to GW emission. The orbital shrinkage rate agrees within error limits with the rate calculated by Damour, Gopakumar and Iyer (2004). At present, we are using the system to test General Relativity in certain strong field regime unexplored so far. This involves constraining a hypothetical fifth force that arises as a modification of General Relativity in some models of dark matter, dark energy and unification theory. Additionally, we confirm that monitoring of the next flare, predicted to peak on July 31, 2019, at noon GMT, should allow us to test, for the first time, the celebrated black hole no-hair theorem for a massive black hole at the 10% level (Valtonen *et al.* 2011).

Acknowledgement

RH acknowledges GACR grant 13-33324S.

References

Babak, S., Petiteau, A., Sesana, A., Brem, P., Rosado, P. A., Taylor, S. R., Lassus, A., Hessels, J. W. T., Bassa, C. G., Burgay, M. *et al.* 2016, *MNRAS*, 455, 1665

Barker, B. M. & O'Connell, R. F. 1979, *Gen.Relativ.Gravit.*, 11, 149
Damour, T., Gopakumar, A., & Iyer, B. R. 2004, *Phys.Rev.D*, 70, 064028
Gopakumar, A., Iyer, B. R., & Iyer, S. 1997, *Phys.Rev.D*, 55, 6030
Graham, A. W. & Scott, N. 2013, *ApJ*, 764, 151
Hudec, R., Basta, M., Pihajoki, P., & Valtonen, M. 2013, *A&A*, 559, A20
Hudec, R. 2017, private communication
Ivanov, P. B., Igumenshchev, I. V., & Novikov, I. D. 1998, *ApJ*, 507, 131
Lehto, H. J. & Valtonen, M. J. 1996, *ApJ*, 460, 207
Liu, J., Zhang, Y., Zhang, H., Sun, Y., & Wang, N. 2012, *A&A*, 540, A67
Nilsson, K., private communication
Pihajoki, P. 2016, *MNRAS*, 457, 1145
Mora, T. & Will, C. M. 2004, *Phys.Rev.D*, 69, 104021
Sillanpää, A., Haarala, S., Valtonen, M. J., Sundelius, B., & Byrd, G. G. 1988, *ApJ*, 325, 628
Sundelius, B., Wahde, M., Lehto, H. J., & Valtonen, M. J. 1996, *ASP-CS*, 110, 99
Sundelius, B., Wahde, M., Lehto, H. J., & Valtonen, M. J. 1997, *ApJ*, 484, 180
Takalo, L. O., Kidger, M., de Diego, J. A., Sillanpää, A., Piirola, V., & Teräsranta, H. 1990, *A&AS*, 83, 459
Valtonen, M. J., Lehto, H. J., Sillanpää, A., Nilsson, K., Mikkola, S., Hudec, R., Basta, M., Teräsranta, H., Haque, S., & Rampadarath 2006, *ApJ*, 646, 36
Valtonen, M. J. 2007, *ApJ*, 659, 1074
Valtonen, M. J., Lehto, H. J., Nilsson, K., Heidt,J., Takalo, L. O., Sillanpää, A., Villforth, C., Kidger, M., Poyner, G., Pursimo, T. *et al.* 2008, *Nature*, 452, 851
Valtonen, M. J., Nilsson, K., Villforth, C., Lehto, H. J., Takalo, L. O., Lindfors, E., Sillanpää, A., Hentunen, V.-P., Mikkola, S., Zola, S. *et al.* 2009, *ApJ*, 698, 781
Valtonen, M. J., Mikkola, S., Merritt, D., Gopakumar, A., Lehto, H. J., Hyvönen, T., Rampadarath, H., Saunders, R., Basta, M., & Hudec, R. 2010, *ApJ*, 709, 725
Valtonen, M. J., Mikkola, S., Lehto, H. J., Gopakumar, A., Hudec, R., & Polednikova, J. 2011, *ApJ*, 742, 22
Valtonen, M. J., Ciprini, S., & Lehto, H. J. 2012, *MNRAS*, 427, 77
Valtonen, M. J. & Pihajoki, P. 2013, *A&A*, 557, 28
Valtonen, M. J., Zola, S., Ciprini, S., Gopakumar, A., Matsumoto, K., Sadakane, K., Kidger, M., Gazeas, K., Nilsson, K., Berdyugin, A. *et al.* 2016, *ApJ*, 819, L37 (OJ287-15/16 Collaboration)
Will, C. M. & Maitra, M. 2017, *Phys.Rev.D*, 95, 064003

Gravitational Wave Astrophysics: Early Results from Gravitational Wave Searches and Electromagnetic Counterparts
Proceedings IAU Symposium No. 338, 2017
G. González & R. Hynes, eds.

doi:10.1017/S1743921318000492

Background Rejection using Convolutional Neural Networks

Adam Zadrożny[1] **and Beata Goźlińska**[2]

[1]Center for Gravitational Wave Astronomy, University of Texas Rio Grande Valley
Cavalry 105, One West University Blvd., Brownsville, Texas 78520, USA
email: `adam.zadrozny@utrgv.edu`

[2] Faculty of Physics, University of Warsaw,
Krakowskie Przedmiecie 26/28, 00-927 Warszawa, Poland
email: `b.gozlinska@student.uw.edu.pl`, `b.gozlinska@gmail.com`

Abstract. The paper presents a proof of concept method of background rejection based on convolutional neural networks (CNN). The method was tested on simulated data and achieved very high accuracy (100%). What is more, method based on CNN is very fast and could be easily applied to wide field surveys. Since early stage results suggest method is very accurate and robust, it could be helpful in creating very low-latency pipelines for EM Follow-up purposes, which will be needed in LIGO-Virgo O3 EM Follow-up.

Keywords. Convolutional Neural Networks, Electromagnetic Follow-up, Background Rejection

1. Introduction and Motivation

For effective searching for electromagnetic counterparts, to gravitational waves (GW) detected by LIGO-Virgo detectors, the most important is to successfully identify a optical transients on images taken by the telescopes. Typically for each event detected by the network of three GW detectors for O2 LIGO-Virgo science run an area of the sky to be searched is around ~100 *sq.deg* (Abbott *et al.* 2017). This is a challenge even for a big field of view telescopes. If transient is identify early enough it gives an opportunity to other telescopes to make a more detailed follow-up and taking spectra. There is a clear need of a fast and very reliable system for distinguishing, real optical transients (OTs) from background events (bogus transients), that could be used for low-latency analysis. Such system could be used for wide deep survey like TOROS (Díaz *et al.* 2017).

2. Proof of concept method

Current methods based on Laplace transform are considered the most commonly used. But those methods not always could distinguish correctly between real and bogus events. Since, in border cases human inspection of images is the final criteria, it seem to be to important to check, if methods based on neural network could be used on top of standard methods. Especially worth considering are convolutional neural networks widely used for image processing. In case of papers describing background rejection methods (van Dokkum 2001, Bailey *et al.* 2008), it is suggested that neural nets (not CNNs) are much worse than other methods (for example Laplace transform). But it has to be noted that those articles (van Dokkum 2001, Bailey *et al.* 2008) considered normal neural nets. It seems that apart from Cherenkov telescopes, neural nets (but not CNNs) (Bernlhr *et al.* 2013) are not used for background rejection. CNN types of networks are used for classifying radio galaxies (Aniyan & Thorat 2017).

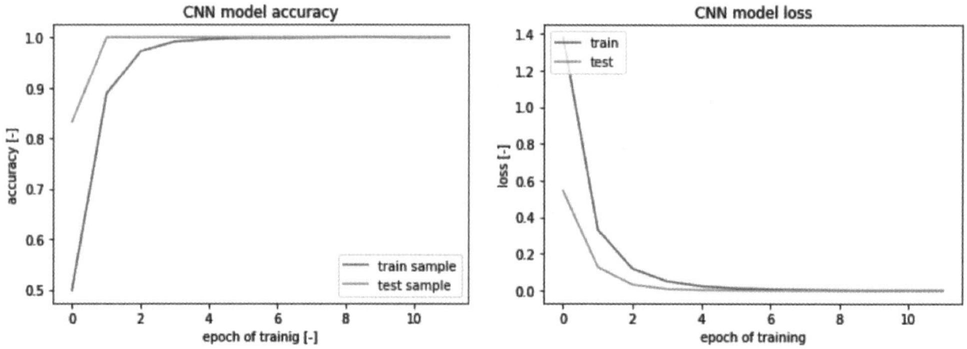

Figure 1. Accuracy (left) and loss (right) of classification done by CNN vs. training epoch.

Structure of CNN. Proof of concept method using CNN was build to classify, if an object is real optical transient or a bogus transient event. All code was written in python 2.7 programming language. For constructing neural network models a Keras package with Theano backend was used. The model of network was constructed using following layers: 2D convolutions along with 2D max polling (2x), flatten (1x), dense (3x).

Training of the network. The network was trained and tested using simulated telescope data. Input for the network was 21 by 21 pixels stamp image of potential optical transient. Image consisted of an object and surrounding background. There was no other source on the stamp image. There were six classes of possible objects: real optical transients, noisy background, hot pixel, cosmic rays, saturation from bright star and artifacts from badly opened shutter. Both train and test samples consisted of 8000 simulated cases divided into six classes. This corresponds to roughly 1333 cases per class. Network was trained using 12 epochs. The trained CNN was able to positively classify all cases, which means that probability of error is less than 1/8000. What is more method turned out to be very fast. One thousand cases was analyzed by CNN in 350 ms on laptop Core i7 processor. The proof of concept shows that based on CNN very accurate and fast methods for background rejection could be build.

Further Work. The developed method has to be tested on real data with injected artificial optical transient. In the next step it would be tested on real data from telescopes of TOROS collaboration. If above tests are successful, the method could be implemented into TOROS data analysis pipeline to make low-latency analysis more accurate for EM Follow-up.

3. Summary

A proof of concept method for rejecting background events transients using Convolutional Neural Network proven to be efficient and reliable. The method was tested on simulated data. The artificial neural network used in the method have reached 100% classification rate of real/bogus transients on 8000 samples. Very high accuracy of this solution suggest that classification methods based on CNN's could be used for low-latency image analysis for LIGO-Virgo's EM Follow-up related observations, and could allow automated EM counterpart discovery and GCN issuance.

The work presented in this paper was done with in scope of NSF#HRD:1242090 research grant.

References

Abbott, B. P. *et al.* 2017, *Phys. Rev. Lett.* 119, 161101
Aniyan, A. K. & K. Thorat, K. 2017, *ApJS* 230, 20
Bailey, S. *et al.* 2008, *Astronomische Nachrichten* 329 292
Bernlhr, K. *et al.* 2013, *Astroparticle Physics* 43 171
Díaz, M. C. *et al.* 2017, *ApJL* 848, L29
van Dokkum, P. G. 2001, *PASP* 113, 1420

Gravitational Wave Astrophysics: Early Results from Gravitational Wave Searches and Electromagnetic Counterparts
Proceedings IAU Symposium No. 338, 2017
G. González & R. Hynes, eds.

doi:10.1017/S1743921318000182

Merging massive black holes the right place and the right time

Astrid Lamberts
Theoretical Astrophysics, California Institute of Technology,
Pasadena, CA, 91125, USA
email: lamberts@caltech.edu

Abstract. The LIGO/Virgo detections of gravitational waves from merging black holes of $\simeq 30$ solar mass suggest progenitor stars of low metallicity ($Z/Z_\odot \lesssim 0.3$). In this talk I will provide constrains on where the progenitors of GW150914 and GW170104 may have formed, based on advanced models of galaxy formation and evolution combined with binary population synthesis models. First I will combine estimates of galaxy properties (star-forming gas metallicity, star formation rate and merger rate) across cosmic time to predict the low redshift BBH merger rate as a function of present day host galaxy mass, formation redshift of the progenitor system and different progenitor metallicities. I will show that the signal is dominated by binaries formed at the peak of star formation in massive galaxies with and binaries formed recently in dwarf galaxies. Then, I will present what very high resolution hydrodynamic simulations of different galaxy types can learn us about their black hole populations.

Keywords. galaxies:abundances, stellar content; stars:binaries, black holes, evolution; gravitational waves

---

1. Introduction

The detection of gravitational waves (GW) from merging black holes (BH) and neutron stars has opened a new window on our Universe (LIGO/Virgo (2016), LIGO/Virgo (2017)). More specifically, it has revived the study of massive stars and their binary interactions (Belczynski *et al.* (2016), Eldridge & Stanway (2016), Spera & Mapelli (2017)). While there is a global understanding of binary evolution for field binaries, the exact properties of their mass loss, mass transfer and supernova mechanisms are still unclear (Belczynski *et al.* (2010), Dominik *et al.* (2013)). The high masses of the first GW detection, (30 $M_\odot$ BHs for GW150914) were unexpected (but see Belczynski *et al.* (2010)). These high masses suggest low-metallicity progenitors, where mass loss due to winds is reduced. We have limited information on low-metallicity massive stars as massive stars are short-lived and stars recently formed in the Milky Way have a metallicity at least as high as the Sun. In this talk, we determine the conditions where the progenitors of GW150914 may have formed, based on a semi-analytic formalism of galaxy evolution and a binary population synthesis model. We determine the formation rate of GW150914-like progenitors based on the mass of the host galaxy and the formation time of the progenitor. We refer the reader to Lamberts *et al.* (2016) for a more detailed description of the method. We then present a preliminary analysis of the merger rates predicted by cosmological simulations.

2. Forming low-metallicity stars

The first step is to determine the conditions for low-metallicity star formation in the universe. Globally, the metallicity in the Universe increases as successive generations of

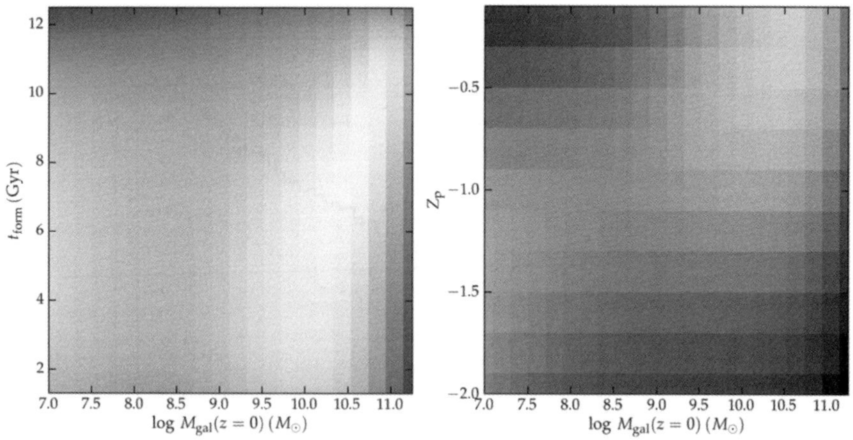

Figure 1. Low-metallicity star formation rate (arbitrary units) as a function of galaxy mass and lookback time to formation (left) and metallicity (right).

stars progressively enrich the star-forming gas. At a given epoch galaxies follow the mass-metallicity relation (Kewley & Ellison (2008)), with more massive galaxies being more metal rich. For each present-day galaxy mass, we determine the mass of the corresponding dark matter halo using abundance matching by Behroozi *et al.* (2013). Based on the model by Behroozi *et al.* (2013) we then determine the amount of star formation in each halo in 10 Myr time bins. Based on the redshift-dependent mass-metallicity relation computed by Ma *et al.* (2016), we derive the stellar mass formed in 11 metallicity bins between $Z = 0.01Z_\odot$ and $Z = 1Z_\odot$, where $Z_\odot \equiv 0.02$. We include scatter between different galaxies of $\sigma = 0.1$ dex, based on observations by Tremonti *et al.* (2004). We also include $\sigma = 0.2$ scatter of the metallicity within galaxies, according to Berg *et al.* (2013). Fig. 1 shows the normalized star formation rate as a function of present-day galaxy mass and lookback time to formation (left) and metallicity (right). The figure only shows subsolar metallicity star-formation, which is relevant to binary BH formation. However, particularly in massive galaxies, a large fraction of the stars form at higher metallicities. We find a slightly bimodal distribution, with most of the low-metallicity stars forming in massive (Milky-Way like) galaxies around the peak of cosmic star-formation (between 6 and 10 Gyrs ago). More recent low-metallicity star formation also occurs in dwarf galaxies, typically at metallicity below 10 per cent of Solar. In the next section, we will highlight how metallicity affects binary BH formation and merger rates and show which low-metallicity stars shown in Fig. 1 mostly contribute to GW150914-like events observed now.

3. Forming binary black hole merger progenitors

We use the BSE code (Hurley *et al.* (2002)) to compute the delay time distribution of binary BH mergers with respect to the formation of the stellar progenitor. We neglect BH mergers from globular clusters (Rodriguez *et al.* (2015)) and consider a single set of standard assumptions on binary evolution. We have updated BSE to account for improvement models of stellar winds (Belczynski *et al.* (2010)), remnant masses (Belczynski *et al.* (2008)) and BH kicks due to supernova explosions (Dominik *et al.* (2013)). Binaries undergoing common-enveloppe mass-transfer during the Hertzprung gap merge as stars. At later stages, we set the common enveloppe efficiency to unity and assume that during Roche lobe overflow, only half of the mass is accreted by the companion.

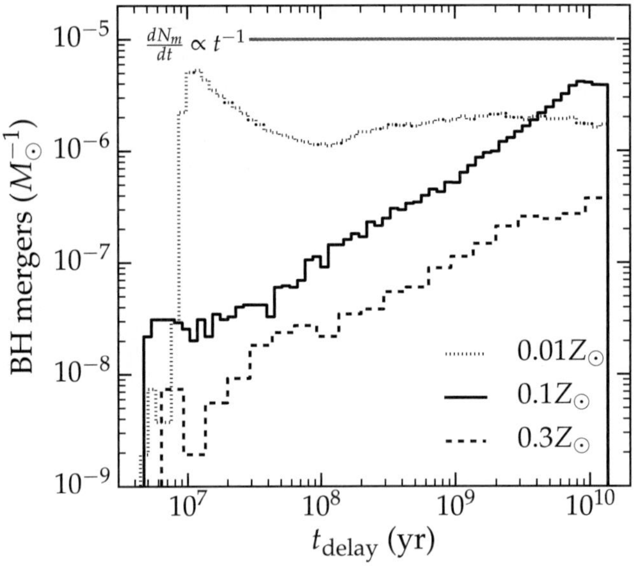

Figure 2. Delay time distribution for massive binary black hole mergers with progenitors of different metallicities. Taken from Lamberts *et al.* (2016).

For each metallicity bin, we model 2.5×10^6 binaries with primary masses between 25 and 150 $M_\odot$ following the Kroupa (2001) initial mass-function. The initial period distributions and mass ratios are derived from observations by Sana *et al.* (2012) and we use thermal distribution for the initial eccentricity. Fig. 2 shows the BH mergers with total mass above 40 $M_\odot$ (per unit solar mass) for $Z = 0.01, 0.1, 0.3Z_\odot$ according to our models. The lowest metallicity stars produce the most massive BH, as their mass-loss is limited and early mergers. At later times, more metal-rich stars contribute more. Stars with metallicity above 0.3 $Z_\odot$ produce 10 times less mergers of these high masses, owing to the strong mass loss of the progenitor stars.

We combine our semi-analytic model for low-metallicity star-formation (Fig. 1) and the metallicity-dependent delay-time distribution (Fig. 2) to determine where and when the progenitor of GW150914 most likely formed. For all stars formed, at all timesteps, we count the resulting BH mergers occurring between $z = 0.1$ and the present day. The conditions of formation of the progenitors are shown in Fig. 3. The figure has the same axes as Fig. 1 and the colors can be directly compared. It shows that the bimodality in low-metallicity star formation is enhanced. Roughly half of the progenitors of GW150914-like systems stem from Milky-Way like galaxies ($M \simeq 10^{11} M_\odot$) and were formed 6 to 10 Gyrs ago, at roughly 10 per cent of Solar metallicity. Later star formation in massive galaxies has too high metallicity to significantly contribute to BH merger progenitors. Recent star formation in dwarf galaxies also significantly contributes to BH mergers, stemming from progenitors with metallicities below 10 per cent of Solar. Because of their low-metallicity, galaxies below $10^8 M_\odot$ overproduce BH mergers in comparison with their contribution to the global star formation. These galaxies are faint, and unobservable at high redshift and BH mergers may be the only way to infer some of their properties. The side panel of the left figure shows the formation time of the progenitors of massive BH mergers, which is mostly uniform over the past 7 Gyrs. This differs from the bimodal distribution from Belczynski *et al.* (2016), because of our improved model of low-metallicity star-formation. In the next section we present preliminary results from cosmological simulations.

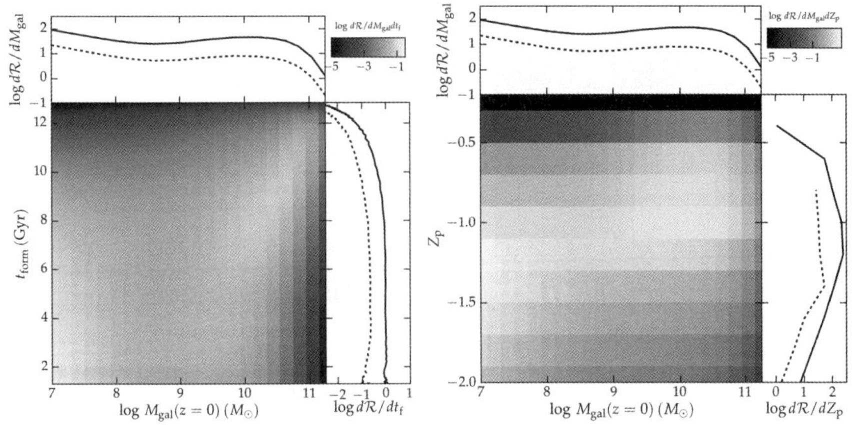

Figure 3. Formation rate of progenitors of massive black hole mergers as a function of galaxy mass and lookback time to formation (left) and metallicity (right). Taken from Lamberts *et al.* (2016).

4. Preliminary results from high-resolution simulations

We have performed a comparable analysis based on hydrodynamical simulations of different galaxies. We use simulations from the Feedback In Realistic Environments project (FIRE ,Hopkins *et al.* (2014)). The cosmological simulations are ran with the GIZMO code (Hopkins (2013)) and include a multiphase model of the interstellar medium and stellar feedback based on a complete stellar evolution model. We specifically use the "Latte" simulation (Wetzel *et al.* (2016)), a cosmological simulation of a Milky-Way like galaxy with a mass resolution of 7070 $M_{\odot}$. The left panel of fig. 4 shows the normalized star-formation rate in a Milky-Way like galaxy. The right panel only shows the stars contributing to BH mergers within the detector horizon during the second observing run. We have directly associated the delay time distributions (Fig. 2) with the star formation in the simulation. We find that mergers come from low-metallicity star-formation and that effectively most of the stars formed in Milky-Way-like galaxies are unable to form merging black holes. As in Lamberts *et al.* (2016), we find that the mergers come from star formation between 6 and 10 Gyrs ago. Fig. 5 shows the total number of mergers as a function of time in different metallicity bins. Again, we confirm the findings from Lamberts *et al.* (2016), with the largest contribution coming from $Z = 0.03 - 0.13 Z_{\odot}$. We find that the number of mergers increases towards redshift $z = 2$ (10Gyrs ago), somewhat following the global star formation history of the galaxy. As the horizon of the detectors increases, we thus expected an increase of the number of detections per unit volume for contributions from massive galaxies.

5. Implications

In this talk, we presented a unique combination between a binary population synthesis model and a complete semi-analytic cosmological model for low-metallicity star formation. We derive the contribution of different galaxies to the present-day merger rate of massive ($> 30 M_{\odot}$) binary black holes. We find that the progenitors of such mergers could have formed during the peak of star formation (between 6 and 10 Gys argo) in a galaxy that would now resemble the Milky Way or more recently (< 6 Gyrs ago) in a dwarf galaxy. Our method solely relies on the strong metallicity dependence of binary black hole formation (and mergers) to determine the *relative* contributions of different galaxies to

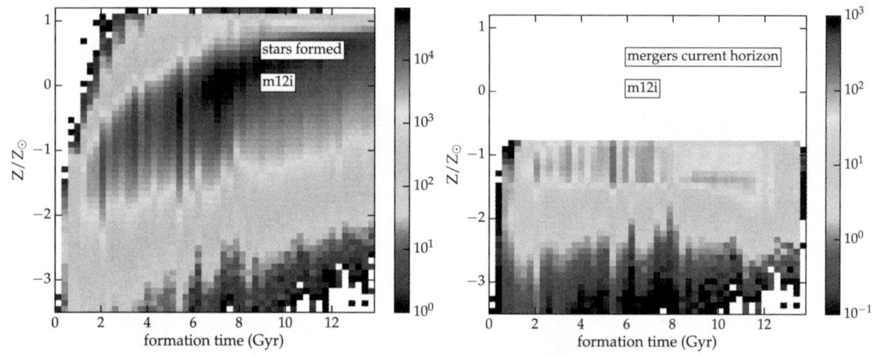

Figure 4. Normalized star formation rate (left) and formation rate of BBH merger progenitors (right) from the Latte simulation.

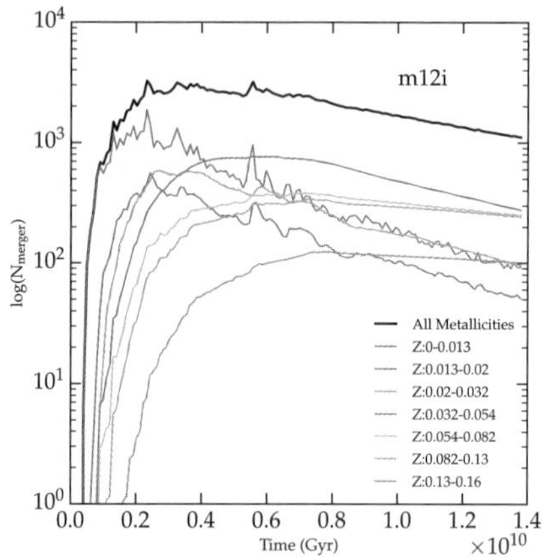

Figure 5. Number of mergers in a Milky-Way like galaxy as a function of time for different metallicities of the progenitor stars. The total number of mergers is shown in the thick black line.

present-day black hole mergers. Although *absolute* merger rate we predict is in agreement with the LIGO/Virgo detection rate (LIGO/Virgo (2016)), we emphasize that it may be revised as our understanding of massive binary evolution improves. We confirm our findings using the same binary evolution model combined with a high-resolution simulation of a Milky-way like galaxy. We find that the merger rate in massive galaxies increases towards redshift $z = 2$ and hundreds of detections are expected when LIGO and Virgo reach their design sensitivity (about $z \simeq 1$). As such, these detections will provide strong constrains on binary evolution models and star formation in high-redshift and/or faint galaxies.

References

Behroozi, P. S., Wechsler, R. H., & Conroy, C. 2013 *ApJ*, 770, 57

Belczynski, K., Holz, D. E., Bulik, T. & O'Shaughnessy, R. 2016 *Nature*, 734, 512

Belczynski, K., Bulik, T., Fryer, C. L., Ruiter, A., Valsecchi, F., Vink, J. S., & Hurley, J. R. 2010 *ApJ*, 714, 1217

Belczynski, K., Dominik, M., Bulik, T., O'Shaughnessy, R. Fryer, C., & Holz, D. E. 2010 *ApJL*, 715, 138

Belczynski, K., Kalogera, V., Rasio, F. A., Taam, R. E., Zezas, A., Bulik, T., Maccarone, T. J., & Ivanova, N. 2008 *ApJ*, 174, 223

Berg, D. A., Skillman, E. D., Garnett, D. R., Croxall, K. V., Marble, A. R., Smith, J. D., Gordon, K., & Kennicutt, Jr., R. C., 2013 *ApJ*, 775, 128

Dominik, M., Belczynski, K., Fryer, C., Holz, D. E., Berti, E., Bulik, T., Mandel, I., & O'Shaughnessy, R. 2013 *ApJ*, 759,52

Eldridge, J. J. & Stanway, E. R. 2016 *MNRAS*, 462, 3302

Hopkins, P. F. 2013 *MNRAS*, 428, 2840

Hopkins, P. F., Kereš, D., Oñorbe, J., Faucher-Giguère, C.-A., Quataert, E., Murray, N., & Bullock, J. S. 2014 *MNRAS*, 445, 581

Hurley, J. R., Tout, C. A., & Pols, O.R. *MNRAS*, 329, 897

Kewley, L. J. & Elisson, S. E. 2008 *ApJ*, 681, 1183

Kroupa, P. 2001 *MNRAS*, 322, 231

Lamberts, A., Garrison-Kimmel, S., Clausen, D., & Hopkins, P. F. . 2016 *MNRASL*, 463, 31

LIGO/Virgo Scientific Collaboration 2016 *ApJL*, 833, 1

LIGO/Virgo Scientific Collaboration 2016 *PRL*, 116, 061102

LIGO/Virgo Scientific Collaboration 2017 *PRL*, 119, 161101

Ma, X., Hopkins, P. F., Faucher-Giguère, C.-A., Zolman, N., Muratov, A. L., Kereš, D., & Quataert, E. 2016 *MNRAS*, 456, 2140

Rodriguez, C. L., Morscher, M., Pattabiraman, B., Chatterjee, S., Haster, C.-J., & Rasio, F. A. 2017 *PRL*, 5, 051101

Sana, H., de Mink, S. E., de Koter, A., Langer, N., Evans, C. J., Gieles, M., Gosset, E., Izzard, R. G., Le Bouquin, J.-B., & Schneider, F. R. N 2012 *Science*, 337, 444

Spera, M. & Mapelli, M. 2017 *MNRAS*, 470, 4739

Tremonti, C. A., Heckman, T. M., Kauffmann, G.& Brinchmann, J., Charlot, S., White, S. D. M., Seibert, M., Peng, E. W., Schlegel, D. J., Uomoto, A., Fukugita, M., & Brinkmann, J. 2004 *ApJ*, 613, 898

Wetzel, A. R., Hopkins, P. F., Kim, J.-h., Faucher-Giguère, C.-A., Kereš, D., & Quataert, E. 2016 *ApjL*, 827, L23

Gravitational Wave Astrophysics: Early Results from Gravitational Wave Searches and Electromagnetic Counterparts
Proceedings IAU Symposium No. 338, 2017
G. González & R. Hynes, eds.

doi:10.1017/S1743921318003617

Supermassive Black Hole Binary Candidates from the Pan-STARRS1 Medium Deep Survey

Tingting Liu[1] **and Suvi Gezari**[1]
[1]Department of Astronomy, University of Maryland, College Park, Maryland 20742, USA
email: tingting@astro.umd.edu

Abstract. We conducted a systematic search for periodically varying quasars, which are predicted manifestations of sub-pc supermassive black hole binaries (SMBHBs), in the Pan-STARRS1 Medium Deep Survey (PS1 MDS). Since the normal variability of quasars can also mimic periodicity over a small number of cycles, we have extended the temporal baseline by monitoring the candidates with the Discovery Channel Telescope and the Las Cumbres Observatory telescopes. We have also adopted a more rigorous method to evaluate the significance of the periodic candidates, by considering in the light curves a "red noise" background modeled as the Damped Random Walk process. While none of the candidates can be resolved by the current pulsar timing arrays (PTAs) as individual gravitational wave sources, the Large Synoptic Survey Telescope is capable of finding more periodic candidates, some of which are likely to be detected by the PTA experiment with the Square Kilometre Array.

Keywords. quasars: individual, surveys, gravitational waves

---

1. Introduction

Supermassive black hole binaries (SMBHBs) should be common products as galaxies grow through mergers in the ΛCDM Universe (e.g. review by Kormendy & Ho 2013). The supermassive black holes (SMBHs) in the centers of the merging galaxies form a pair at < 10 kpc separations and can be observed as a dual active galactic nucleus (AGN) when both SMBHs are actively accreting. The binary orbit further shrinks to < 1 pc by interaction with stars and a gas disk, and gravitational wave radiation eventually drives the SMBHB to coalesce. SMBHBs at close separations are therefore also the loudest expected gravitational wave (GW) sources for the pulsar timing arrays (PTAs) in the nHz-μHz frequency band.

Many theoretical efforts have been dedicated to exploring the *indirect* observational signatures of an SMBHB. (Magneto-) Hydrodynamical simulations of an SMBHB system (e.g. MacFadyen & Milosavljević 2008; Shi *et al.* 2012; Noble *et al.* 2012; D'Orazio *et al.* 2013; Gold *et al.* 2014) find that (1) the binary torque clears and maintains a low-gas density cavity of a radius $r \sim 2a$ (where a is the binary separation) in the circumbinary disk, and material is ushered in through a pair of accretion streams; and (2) the accretion rate onto the binary is strongly modulated on the order of the binary orbital frequency for various mass ratios. Thus, for typical SMBH masses, the SMBHBs should manifest as quasars that periodically vary on the timescale of months to years, assuming the accretion luminosity directly tracks the mass accretion rate. More recently, D'Orazio *et al.* (2015) also proposed a relativistic Doppler boosting model to interpret the variability of the SMBHB candidate PG 1302–102 (first reported by Graham *et al.* 2015a), where the emission from the mini-disk of the secondary black hole is Doppler boosted as it travels at a relativistic speed along the line of sight.

There have been a number of systematic searches for periodic quasars as SMBHB candidates in optical time domain surveys — e.g. Graham *et al.* (2015a) and Graham *et al.* (2015b) in the Catalina Real-time Transient Survey (CRTS); Charisi *et al.* (2016) in the Palomar Transient Factory (PTF); and Liu *et al.* (2015), Liu *et al.* (2016), and Liu *et al.* (in preparation) in the Pan-STARRS1 (Kaiser *et al.* 2010) Medium Deep Survey (PS1 MDS) — and more than 100 SMBHB candidates have been claimed so far.

However, most of those candidates have less than $\sim 2-3$ cycles of variation, making it easy for normal, stochastic variability of AGN and quasars to masquerade as periodic sources (e.g. Vaughan *et al.* 2016). Therefore, after conducting a systematic search in PS1 MDS for periodic quasar candidates (§2), we extend the observational baseline with new imaging data taken at the Discovery Channel Telescope (DCT) and the Las Cumbres Observatory (LCO) network and adopt a more rigorous maximum likelihood method from Zoghbi *et al.* (2013) which takes into account a "red noise" background (§3). We have down-selected a sample of more robust candidates and are able to infer binary parameters from this population of SMBHB candidates. We conclude with discussions of GW detection of SMBHBs by the PTAs in §4.

2. A systematic search for periodic quasars in PS1 MDS

PS1 MDS surveyed 10 circular fields distributed across the sky, each of which is ~ 8 deg^2 in size, from 2009-2014. It observed in the g r i z y filters, the first four of which have been used in our study. Each MD field is chosen such that it overlaps with an existing extragalactic survey field. In our pilot study in the MD09 field (Liu *et al.* 2015; Liu *et al.* 2016), which overlaps with SDSS Stripe 82, we first cross-matched the 40,488 point sources extracted from MDS with a custom catalog of deep stacks in the Canada-France-Hawaii Telescope (CFHT) u band and PS1 g r i z bands (Heinis *et al.* 2016), after applying a magnitude cut at $m < 23$ mag. We converted the magnitudes to the SDSS system and apply the color box from Sesar *et al.* (2007) and selected 670 quasars by color. We then measure the observed variability (defined as the standard deviation σ of the light curve) of each quasar and compare with that of an ensemble of nearby stars of similar brightness. As most stars are not intrinsically variable, any scatter in their light curves should only be due to photometric errors and other systematic effects and therefore establish our sensitive limit in detecting variable sources. We selected 104 variable quasars at the $> 2\sigma$ level.

Using the Lomb-Scargle periodogram (Horne & Baliunas 1986) and taking advantage of PS1 monitoring in multiple filters, we then look for a coherence periodic signal in at least three of the four filters. Among the 77 variable quasars that met this requirement, we also require a minimum signal-to-noise (> 3) and at least 1.5 cycles of variation. The search resulted in 3 periodic candidates from a ~ 5 deg^2 of PS1-CFHT cross-matched sky area (Liu *et al.* 2016).

3. Putting candidates to the test

However, stochastic variability of normal quasars and AGN, characterized by either the Damped Radom Walk model (DRW; Kelly *et al.* 2009) or a power spectral density (PSD) model of a broken power law, can easily mimic a periodic variation over a small number of cycles ($\sim 2-3$), especially with large photometric errors and sparse sampling (Vaughan *et al.* 2016). Therefore, to address the issue of "red noise" contaminating our selected sample, we have extended the baseline of observations using available archival light curves from the SDSS Stripe 82 survey (2000-2008) and have also been monitoring the candidates with new imaging data taken with the Large Monolithic Imager (LMI) at DCT (since 2015 May) and the Spectral imager on the LCO network telescopes (since 2017 April).

Our imaging programs on DCT/LMI and LCO/Spectral are carried out in the SDSS g r i z filters. We perform aperture photometry on the reduced images using `SExtractor` and cross-match the extracted sources with point sources from the SDSS catalog in order to convert from the `SExtractor` instrumental magnitude to an SDSS magnitude. To convert the SDSS magnitudes to the PS1 photometric system and thus directly compare with data from MDS, we calculate synthetic PS1 and SDSS magnitudes by convolving the (redshifted) composite quasar spectrum from Vanden Berk *et al.* (2001) with the respective filter sensitivity curve and then apply the magnitude offset to the LMI or LCO measurements. We then "stitched" the new measurements and SDSS Stripe 82 archival light curve with the PS1 light curve to generate an extended light curve.

We then re-ran our analysis on the extended light curves for the three candidates from MD09. None of them passed the test as persistent periodic sources, further demonstrating the importance of having long-term monitoring for candidates selected based on their apparent periodicity (Liu *et al.* 2016).

We have expanded our analysis to all 10 MD fields in Liu *et al.* (in preparation) and selected 26 periodic candidates from 9,314 color-selected quasars. To more rigorously test their periodicity, in addition to extended baseline monitoring, we have also adopted the maximum likelihood method from Zoghbi *et al.* (2013) to re-evaluate the significance of the periodic signal in the presence of a red noise background. We model the PSD as a DRW process and search for a periodic component, which is represented by a δ-function. By comparing the likelihoods of "DRW only" and "DRW+periodic" models, we determine whether the additional periodic component is justified in both the "PS1-only" and extended light curves. We are therefore able to down-select 5 more robust candidates in which the "DRW+periodic" model is preferred (Fig. 1).

While we have directly measured quantities such as black hole mass, redshift, and variability period and have inferred a binary orbital separation by virtue of Kepler's law, we further demonstrate that it is also possible to infer or constrain other binary parameters such as mass ratio from a population of variability-selected SMBHB candidates. We compute the binary residence times $t_{\rm res}$ for the PS1 MDS candidates for a range of mass ratios $q = 0.01 - 1$ and compare the distribution with that of a population of purely GW-driven SMBHBs. After correcting for observational constraints due to the limited baseline of PS1 MDS, we find that the full sample of 26 SMBHB candidates is poorly matched to the expected $t_{\rm res}$ distribution for all mass ratio cases, while the down-selected sample is well-matched to the expected distribution for $q = 0.3$ (Fig. 2), consistent with the major galaxy merger scenario.

SMBHB candidates selected from an optical time domain survey should also be followed up with multi-wavelength observations in order to further study their observational properties; they can also provide independent evidence that could verify the binary hypothesis. In the UV and optical wavelengths, for example, Roedig *et al.* (2014) predicted that the cavity in the circumbinary disk should produce a spectral "notch", which can be probed in the SED or optical/UV spectrum of the quasar. They have also predicted that due to the accretion stream shock-heating the edge of the mini-disk, it would produce a hot spot that corresponds to a hard X-ray excess at $\sim$ 100 keV. For SMBHB candidates at high redshifts (z$\gtrsim$ 1), this unique signature is within the spectral window of *NuSTAR* (3 – 79 keV).

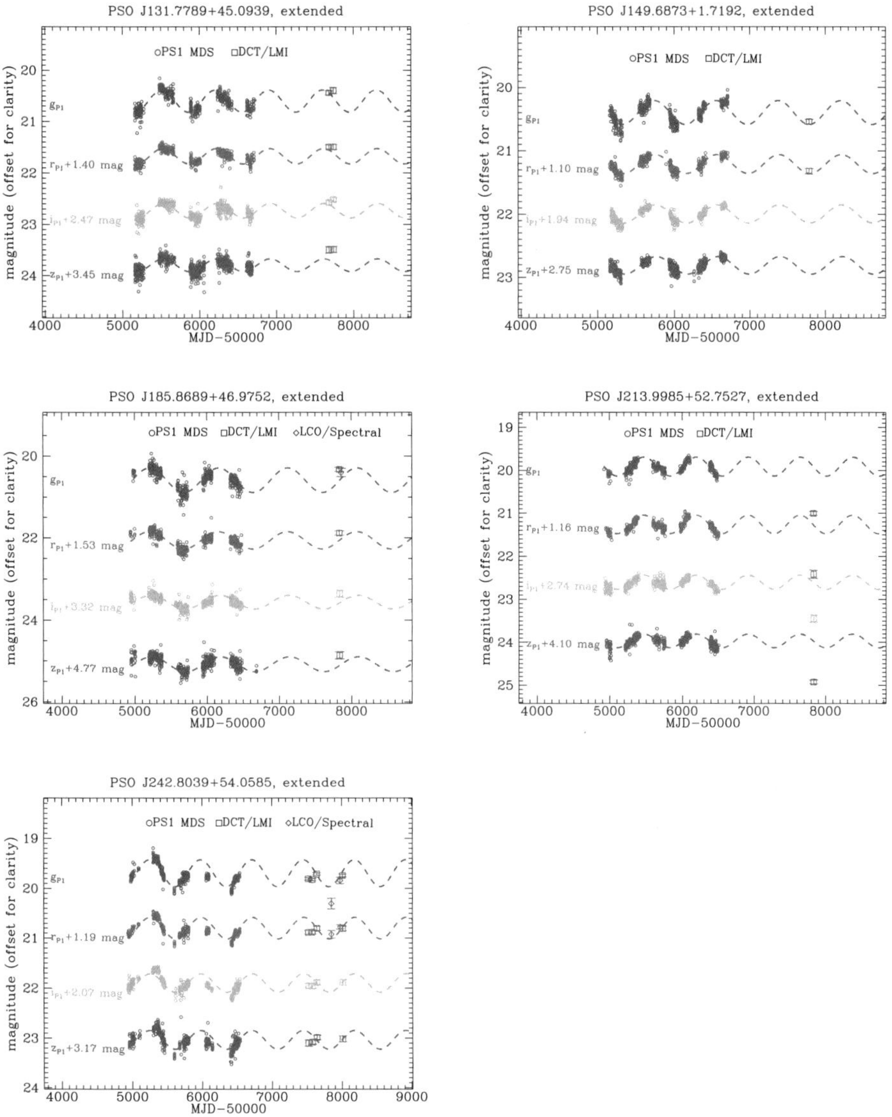

Figure 1. Five "gold" candidates from PS1 MDS in which a periodic component is present after including a DRW background Liu *et al.* (in preparation). Their g r i z band light curves from MDS (circles; photometric error bars have been omitted for clarity) are shown with new monitoring data taken with DCT/LMI (squares) and LCO/Spectral (diamonds) between 2016 May (approximately MJD 57500) and 2017 September (approximately MJD 58000). Sinusoids are fitted to the "PS1-only" light curves and have been extrapolated to guide the eye.

4. SMBHBs as low-frequency GW sources for the PTAs

The search for periodic quasars in large time domain surveys can also open up the study of them as electromagnetic counterparts of nanohertz-frequency GW sources. While none of the candidates from PS1 MDS can yet be resolved by the current PTAs based on our estimation of their binary parameters, we predict based on our benchmark study with MDS that $\sim 20,000$ periodic quasar candidates of more than 3 cycles could be detected by the Large Synoptic Survey (LSST), ~ 100 times more than the total number of currently

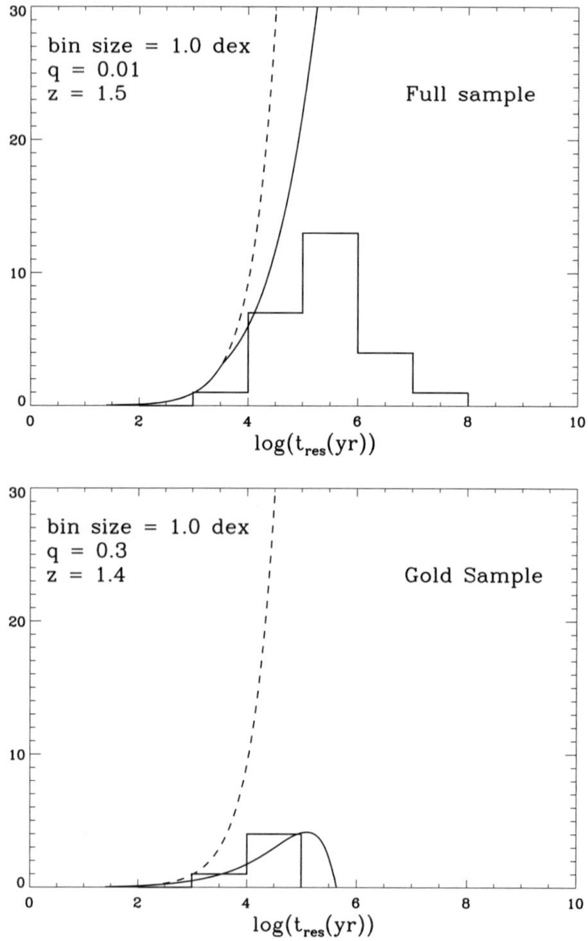

Figure 2. We calculate the expected distribution of residence times ($t_{\rm res}$) for a population of GW-driven SMBHBs (dashed curves) and compare the distribution after taking into account observational constraints (solid curves) with that of the SMBHB candidates selected from PS1 MDS (histograms). The Full sample of 26 candidates is poorly matched to the expected distribution even for an extreme mass ratio q = 0.01 (top panel), while the Gold sample of five selected after our re-analysis is in significantly better agreement with the expected distribution, if $q = 0.3$ is adopted (bottom panel) Liu *et al.* (in preparation).

claimed candidates Liu *et al.* (in preparation). Scheduled to operate for 10 years, LSST will also be able to detect periodic candidates with more cycles, thus decreasing the false alarm rate due to normal quasar variability.

Adopting the probability densities of the binary parameters of the first likely GW sources for the International Pulsar Timing Array (IPTA) and Phases 1 and 2 of the Square Kilometre Array (SKA1 and SKA2) from Rosado *et al.* (2015), we draw realizations of SMBHBs and compare with the parameter space that CRTS, PS1 MDS, or LSST is sensitive to, adopting a minimum of 3 cycles of periodic variation. While none of the current candidates from CRTS or PS1 MDS can be resolved by IPTA as individual GW sources, LSST can probe part of the GW strain-frequency parameter space accessible by SKA1 and SKA2 and thus is likely to yield the first individual GW sources for detection by the SKA (Fig. 3).

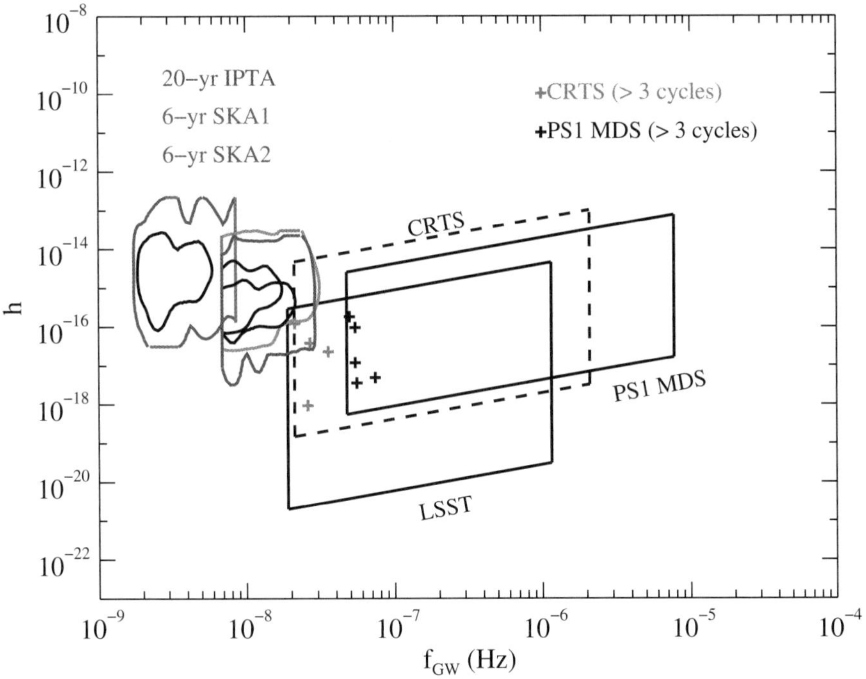

Figure 3. We show the GW frequency (f_{GW}) and strain amplitude (h) of the SMBHB candidates from CRTS (Graham *et al.* 2015b) and PS1 MDS Liu *et al.* (in preparation) that have more than 3 cycles of variation (grey and black crosses, respectively), as well as the f_{GW}-h parameter space that systematic searches in CRTS, MDS, and LSST are sensitive to (dashes, solid, and dotted boxes, respectively). LSST is capable of yielding SMBHB candidates that are the most likely first GW sources detectable by the SKA1 or SKA2 array.

Acknowledgements

TL's attendance at the symposium was partially supported by an IAU travel grant.

References

Charisi, M., Bartos, I., Haiman, Z., *et al.* 2016, *MNRAS*, 463, 2145
D'Orazio, D. J., Haiman, Z., & MacFadyen, A. 2013, *MNRAS*, 436, 2997
D'Orazio, D. J., Haiman, Z., & Schiminovich, D. 2015, *Nature*, 525, 351
Graham, M. J., Djorgovski, S. G., Stern, D., *et al.* 2015, *Nature*, 518, 74
Graham, M. J., Djorgovski, S. G., Stern, D., *et al.* 2015, *MNRAS*, 453, 1562
Gold, R., Paschalidis, V., Etienne, Z. B., Shapiro, S. L., & Pfeiffer, H. P. 2014, *Phys. Rev. D*, 89, 064060
Heinis, S., Gezari, S., Kumar, S., *et al.* 2016, *ApJ*, 826, 62
Horne, J. H. & Baliunas, S. L. 1986, *ApJ*, 302, 757
Kelly, B. C., Bechtold, J., & Siemiginowska, A. 2009, *ApJ*, 698, 895
Kormendy, J. & Ho, L. C. 2013, *ARAA*, 51, 511
Liu, T., Gezari, S., Heinis, S., *et al.* 2015, *ApJ*, 803, L16
Liu, T., Gezari, S., Burgett, W., *et al.* 2016, *ApJ*, 833, 6
Liu, T., Gezari, S. *et al.*, in preparation.
MacFadyen, A. I. & Milosavljević, M. 2008, *ApJ*, 672, 83
Roedig, C., Krolik, J. H., & Miller, M. C. 2014, *ApJ*, 785, 115
Rosado, P. A., Sesana, A., & Gair, J. 2015, *MNRAS*, 451, 2417

Sesar, B., Ivezić, Ž., Lupton, R. H., *et al.* 2007, *AJ*, 134, 2236
Shi, J.-M., Krolik, J. H., Lubow, S. H., & Hawley, J. F. 2012, *ApJ*, 749, 118
Noble, S. C., Mundim, B. C., Nakano, H., *et al.* 2012, *ApJ*, 755, 51
Vanden Berk, D. E., Richards, G. T., Bauer, A., *et al.* 2001, *AJ*, 122, 549
Zoghbi, A., Reynolds, C., & Cackett, E. M. 2013, *ApJ*, 777, 24

Gravitational Wave Astrophysics: Early Results from Gravitational Wave Searches and Electromagnetic Counterparts
Proceedings IAU Symposium No. 338, 2017
G. González & R. Hynes, eds.

doi:10.1017/S1743921318003629

GW170817: *Swift* UV detection of a blue kilonova, and improving the search in O3

Aaron Tohuvavohu[1], Jamie A. Kennea[2] and the *Swift* gravitational wave follow-up group

Department of Astronomy and Astrophysics, The Pennsylvania State University, University Park, PA 16802, USA
email: [1] aaronb@swift.psu.edu [2] kennea@swift.psu.edu

Abstract. *Swift*'s rapid slewing, flexible planning, and multi-wavelength instruments make it the most capable space-based follow-up engine for finding poorly localized sources. During O1 and O2 *Swift* successfully tiled hundreds of square-degrees of sky in the LVC localization regions, searching for, and identifying, possible X-ray and UV/O transients in the field. *Swift* made important contributions to the discovery and characterization of the kilonova AT 2017gfo, discovering the UV emission and providing the deepest X-ray upper limits in the first 24 hours after the trigger, strongly constraining the dynamics and geometry of the counterpart. *Swift* tiled 92% of the galaxy convolved error region down to average X-ray flux sensitivities of 10^{-12} erg cm^{-2} s^{-1}, significantly increasing our confidence that AT 2017gfo is indeed the counterpart to GW 170817 and sGRB 170817. However, there remains significant room for improvement of *Swift's* follow-up in preparation for O3. This will take the form of both revised observation strategy based on detailed analysis of the results from O2, and significant changes to *Swift*'s operational capabilities. These improvements are necessary both for maximizing the likelihood that *Swift* finds a counterpart, and minimizing the impact that follow-up activities have on other *Swift* science priorities. We outline areas of improvement to the observing strategy itself for optimal tiling of the LVC localization regions. We also discuss ongoing work on operational upgrades that will decrease latency in our response time, and minimize impact on pre-planned observations, while maintaining spacecraft health and safety.

---

1. Introduction

The NASA *Swift* Mission (Gehrels *et al.* 2004) is a mid-sized Explorer class satellite designed to study Gamma-Ray Bursts (GRBs). *Swift* has partnered with the LIGO-VIRGO collaboration (LVC) in performing a follow-up program to search for electromagnetic (EM) counterparts to gravitational wave (GW) events since the 2009-2010 science run (Evans *et al.* 2012). The general scheme is a tiling campaign that uses *Swift*'s two narrow field-of-view instruments, the X-ray Telescope (XRT) and UV/Optical Telescope (UVOT), to target high probability fields within the gravitational wave localization region. The *Swift* spacecraft is unique in its ability to carry out tiling campaigns of this type, because of its fast-response and extremely flexible scheduling and commanding, and its singularly rapid slewing. This tiling capability was developed with LIGO tiling in mind, but was extensively tested both by observing simulated LIGO triggers from O1, but also, starting in June 2016, for the *Swift* SMC Survey (S-CUBED; Kennea *et al.*, *in prep.*), a weekly survey covering the Small Magellanic Cloud in 142 pointings of 60 seconds each. Lessons learned in safely performing these rapid re-pointing maneuvers with the spacecraft, building the complex schedules to facilitate the campaigns, and understanding the necessary instrument modes to be able to perform useful science in such brief snapshots were developed by the *Swift* team and brought to bear on the goal of rapidly searching the gravitational wave localization regions. During the post-detection

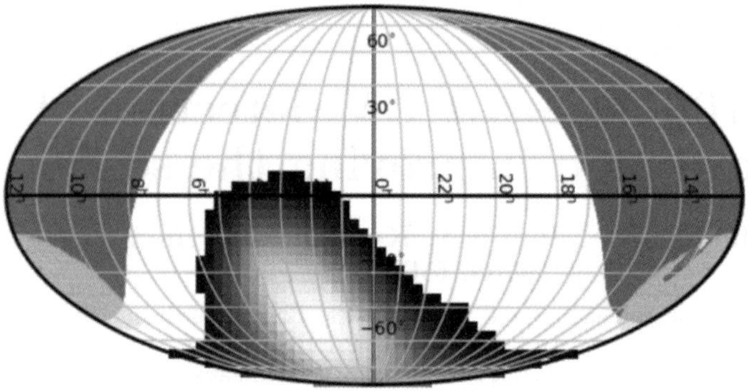

Figure 1. The BAT FoV at the time of the GW/sGRB trigger. The black/red/white region shows the BAT FoV, with the color scale indicating the partial coding fraction. Everything occulted by the Earth from the perspective of *Swift* is shaded green. The blue and yellow regions are the GW and *Fermi* GBM 90% confidence localization regions respectively.

era (O2) *Swift* has responded to many LVC triggers in this capacity, searching hundreds of square-degrees of galaxy convolved probability for a possible EM counterpart using the techniques described by Evans *et al.*(2016).

Because all of the LVC triggers in O1, and most of those in O2, were BBH events, the follow-up campaigns that *Swift* performed put strong limits on the X-ray emission from these events, and were consistent with the expectation that BBH mergers would be dark, but were most useful as test runs to ensure that *Swift*'s response would be optimal when a gravitational wave signal from a system with an expected EM counterpart finally was detected. Results from these campaigns were used to help improve the tiling strategy itself, reduce the response time of the observatory, and test the automated data processing and transient identification pipelines.

2. *Swift*'s response to GW/sGRB 170817

At 12:41:04 UT the LVC registered a gravitational wave signal whose strain waveforms were consistent with a binary neutron star system (Abbott *et al.* 2017). Approximately 2 seconds later *Fermi*'s Gamma-Ray Burst Monitor (GBM) triggered on a short (~ 2 seconds duration) gamma-ray signal consistent with the GW localization. At this time, *Swift* was at a location in its orbit such that the GRB and GW localization regions were occulted by the Earth limb (Figure 1), and so the relevant region of the sky was not visible to *Swift*'s Burst Alert Telescope (BAT).

News of a GW signal temporally coincident with a *Fermi* GBM trigger was disseminated to the community via a private GCN circular at 13:21 UT (LVC, GCN 21505). Upon receipt, the *Swift* team immediately triggered a follow-up campaign. *Swift*'s narrow-field instruments began performing settled observations in search of an EM counterpart at 13:37 UT, 16 minutes after notification. *Swift* was the first telescope to begin pointed observations in response to the trigger. At this time, the best localization available came from GBM (90% containment area of ~ 1626 deg^2) as the multiple detector LIGO skymap took several hours to produce because of a glitch in the LIGO-Livingston data. For this reason, *Swift*'s earliest observations were a 37-point tiling covering the ~ 3.7 deg^2 at the center of the *Fermi* GBM localization region (see Fig. 2).

At 17:54 UT a three-detector sky-map with a 90% containment area of 33.6deg^2 became available from the LVC. This sky-map was convolved with the GWGC catalog, as described in Evans *et al.* 2016. *Swift* promptly began a tiling campaign targeting these

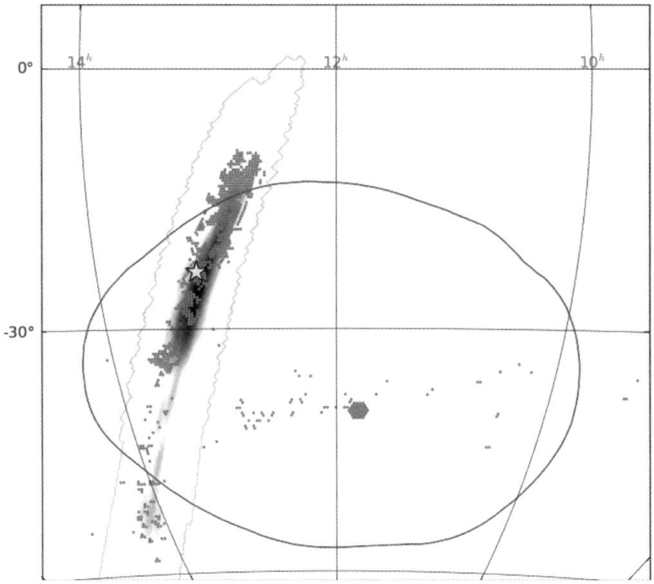

Figure 2. Sky-map of *Swift* observations in equatorial coordinates. Each red pixel is a single XRT observation of 120 s. The dark grey region shows the extent of the final, 3-detector GW localization. The location of AT 2017gfo is marked with a yellow star. The early 37-point tiling can be seen at the center of the blue *Fermi* GBM 90% confidence region. The widely scattered fields were from the first iteration of the observing plan, which was based on the single-detector GW sky-map available at the time.

galaxies, with 120 s of exposure per tile. In this campaign *Swift* observed 92% of the galaxy convolved localization region over the course of 744 individual pointings, finding no evidence for new bright X-ray transients in the field down to a flux upper limit of $f_X \geqslant 10^{-12}$ erg cm^{-2} s^{-1}. Many UV/optical transients were found but none were deemed plausible counterpart candidates.

In order to determine useful constraints from *Swift*'s non-detection in the wide-field search a simulation of 10,000 sGRB afterglows based on *Swift*'s flux-limited sample (D'Avanzo *et al.* 2014) was performed. These light curves were rescaled to the distance derived from the GW signal, and positions were drawn at random from the 3-D 'bayestar-HLV' sky-map (LVC, GCN 21513). We then determined which of the simulated events were in galaxies observed by XRT as part of our tiling campaign, and if the source flux would have been detectable by *Swift* at the time of the observation. It was found that in 65% of cases the source would have been detected. However, sGRB 170817 was quite under-luminous compared to the *Swift* sample. Re-scaling the predicted X-ray flux for the simulated sGRB afterglows down by a factor of 10^3, we find that were there such an afterglow *Swift* would have detected it 11% of the time.

At 01:05 UT on August 18, 2017 the One-Meter Two-Hemisphere (1M2H) team reported the detection of a possible counterpart associated with the galaxy NGC 4993 (Coulter *et al.* 2017), later designated AT 2017gfo. At the time, the nature of this candidate, and whether it was indeed connected to GW/sGRB 170817, was unclear. *Swift* continued its wide-field tiling campaign, but regularly interrupted the tiling to slew to the field of NGC 4993, with the first observation beginning $\sim$ 2 hours after the announcement.

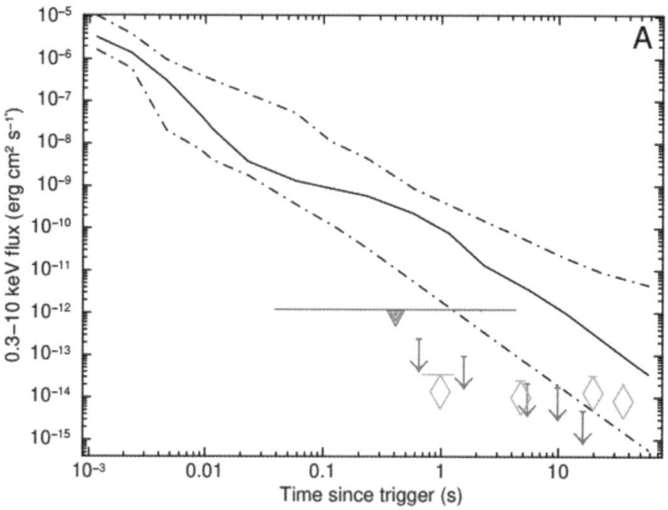

Figure 3. The median behavior and 25-75% distribution of sGRB afterglows, scaled to 40 Mpc, are shown by the solid and dashed lines respectively. The blue line with the triangle shows the time-span over which the tiling observations were taken, and the typical flux sensitivity achieved per tile. The red arrows and grey diamonds show X-ray flux upper limits on emission from AT 2017gfo from *Swift* and NuSTAR observations respectively.

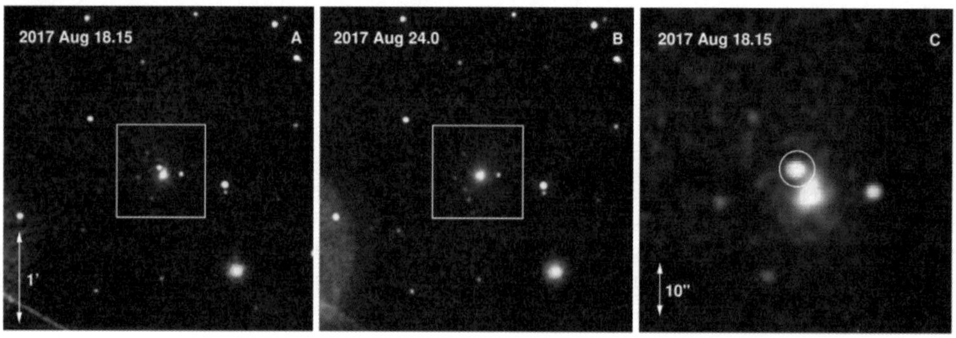

Figure 4. *Swift* UVOT image of the field of NGC 4993. Panel A shows the first epoch, in which bright UV emission is clearly seen. Panel B demonstrates the rapid fading at blue wavelengths. Panel C shows the first epoch zoomed in with AT 2017gfo circled. The *u, uvw1, uvm2* filters are assigned to the red, green and blue channels respectively.

3. Observations of AT 2017gfo

Swift performed the first observation of AT 2017gfo at 03:34 UT on August 18, 2017. In the first exposure *Swift* detected the UV emission from this source, and provided the earliest sensitive X-ray upper limits. AT 2017gfo was detected as a bright and fading UV source with an initial magnitude of $u = 18.9^{+0.09}_{-0.08}$ mag (AB). Figure 5 shows the *Swift* UV and optical light curves, plotted alongside optical and near-infrared photometry from Pan-STARRS (Smartt *et al.* 2017). The rapid fading at UV wavelengths is clearly evident, in contrast to the optical and near-infrared emission which remained detectable for over a week (Kasliwal *et al.* 2017). No X-ray source was found in this first observation with a 3-σ (0.3 - 10 keV) upper limit of 5.5×10^{-3} cts s^{-1}. Assuming an energy conversion typical for GRB afterglows and the distance to NGC 4993 of $\sim$ 40 Mpc, this gives $L_x < 4.2 \times 10^{40}$ erg s^{-1} at 0.6 days after the merger. *Swift* continued monitoring this source with both XRT and UVOT over the 15 days between discovery and when the

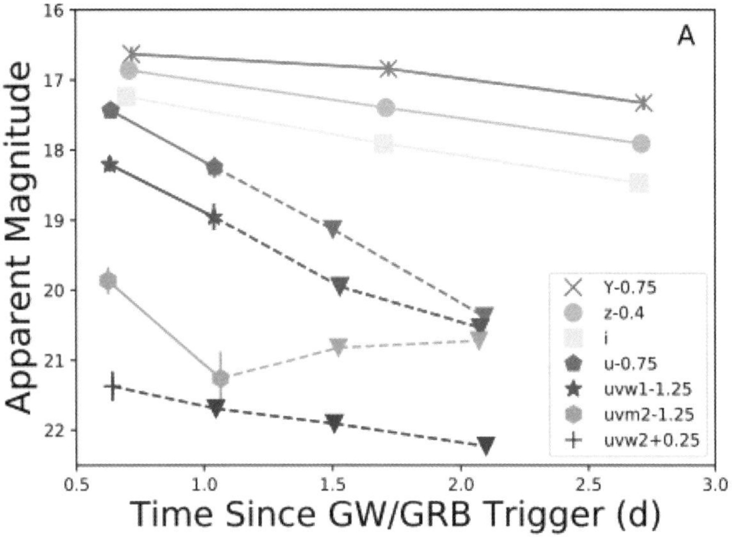

Figure 5. *Swift* UVOT light curve of AT 2017gfo. Upper limits are shown by inverted triangles. The *i,z,y* bands are from Pan-STARRS and are included for the purpose of comparison. All data are corrected for host galaxy contamination.

source became too close to the Sun to observe on September 2, 2017. In this time *Swift* collected 172 Ks of observations, in which no X-ray source was detected. Summing all of the X-ray data gives a 3-σ upper limit of 2.8×10^{-4} cts s^{-1}. With the same assumptions as above this gives $L_x < 2.1 \times 10^{39}$ ergs s^{-1}.

The absence of X-rays at early times, as well as the early UV emission allow independent constraints on the nature of the counterpart. As shown in Figure 3, the lack of X-rays from AT 2017gfo make its interpretation as an on-axis sGRB afterglow a tenuous one. While there are a few sGRBs in the sample with afterglows that fade extremely quickly, and *Swift* would therefore have missed, the vast majority of the sGRB afterglow population would have been detectable. Using the analytic afterglow formulation for synchrotron emission (Evans *et al.* 2017), we can derive physical constraints from the X-ray upper limits. For on-axis viewing the X-ray non-detection independently limits the amount of energy coupled to relativistic ejecta to less than 10^{50} erg, assuming isotropic emission. The X-ray observations can also help inform our understanding of the origin of the UV emission. The optical-to-X-ray spectral index $\beta_{OX} \geqslant 1.6$ at 0.6 days after the merger is far outside the norm for GRB afterglows (van der Horst *et al.* 2009).

The UV/optical SED further discredits the interpretation of the UV counterpart as the emission from a typical sGRB afterglow. The SED can be fit well with a blackbody function with $T_{BB} = 7300 \pm 200$ K at 0.6 days after the merger, with rapid cooling to $T_{BB} = 6400 \pm 200$ K at 1.0 days after the merger. The blackbody model yields an average fit statistic of $\chi^2 = 2.2$ (5 degrees of freedom). For comparison, if the origin of the UV counterpart is from synchrotron afterglow emission, one would expect a power-law model of the form $f_\lambda \propto \lambda^{-\alpha}$ to describe the SED well. At both 0.6 and 1.0 days after the merger we find $\alpha \sim 1.0$. However, the fit quality is extremely poor, with an average fit statistic of $\chi^2 = 98.15$ (5 degrees of freedom). As such, the UV emission must have its origin in a different physical process than an on-axis GRB afterglow. We find that an off-axis orientation of $\approx 30°$ fits the early absence of X-rays along with the late-time Chandra detection well (see Figure 6). Likewise, the presence of UV emission independently constrains the orientation from the other end as sight lines in the plane of

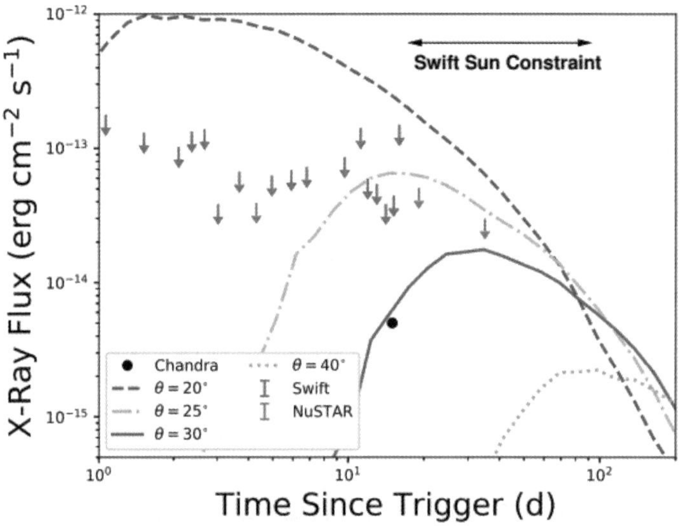

Figure 6. Simulated X-ray afterglow lightcurves for a typical GRB, at different viewing angles. Here $E_{AG} = 2 \times 10^{51}$ erg, $n_0 = 5 \times 10^{-3}$ cm^{-3} and θ_{jet} =0.2 radians.

the merger are expected to have significantly reduced UV emission due to the presence of lanthanide-rich dynamical ejecta. Further implications of the UV emission as well as modeling that shows broad consistency between the early UVOT observations and kilonova models can be found in Evans *et al.* 2017.

4. Future Improvements

While *Swift*'s contributions to the discovery of AT 2017gfo were important and largely successful, there remains room for improvement for future LIGO/Virgo runs such as the upcoming O3. *Swift* is the only satellite observatory capable of performing a fast-response targeted tiling campaign, and will remain alone in this role for O3. In addition, depending on the time at which the GW event is detected, *Swift*'s 96 min orbit means that fields become observable in $<$ 1 hour from detection, whereas ground-based telescopes may have to wait many hours for optimal observing conditions. For this reason the *Swift* team is investing considerable effort into improving response time, tiling strategy, transient detection pipelines, scheduling, and further study into maintaining spacecraft health and safety while performing these campaigns.

4.1. *Scheduling*

Swift has the most complex and demanding scheduling of any satellite observatory, performing on average $\sim$ 120 individual observations per day (compared to a handful per day for most other missions) while juggling a host of astronautical and scientific constraints. Over the lifetime of the *Swift* mission various planning and scheduling tools have been developed to automate as much of this task as possible. However, the construction of optimal schedules has remained the domain of human Science Planners, who outperform the automated tools by significant margins. Unfortunately, it takes a Science Planner approximately 5 hours, on average, to construct an optimal observing plan for one 24-hour period. For rapid response campaigns such as constructing a plan for large-scale LIGO tiling, this is simply too slow. Because of this, the current automated pipeline that constructs the rapid-response tiling campaigns, and associated schedules, to send up to the spacecraft produces plans that are scientifically sub-optimal. In order to rapidly construct these schedules, the LIGO tiling observations are merged with current pre-planned observations for the day, bumping those observations that conflict with LIGO tile observations. In order to fix any issues with the plan, further science targets are dropped in favor of a target from a large pool of 'fill-in' targets, typically targets for which observations are not time critical, to prioritize the health-and-safety of *Swift*.

What this means is that even those pre-approved targets that have locations on the sky that are not at all in conflict with the GW localization region are often bumped out of the schedule, and the spacecraft spends a significant amount of time observing targets that are not of the highest scientific merit. While this was deemed tolerable during O1 and O2, given the expected increased rate of GW triggers in O3 consistently dropping *Swift*'s pre-planned science becomes unacceptable. As such, significant effort has been invested in building a fast, effective, and optimal automated planning pipeline that can outperform humans. This project is described in (Tohuvavohu 2017) and will have the added benefit of increasing the efficiency of the observatory in general, and significantly adding to its already prodigious fast-response capabilities.

4.2. *Commanding*

The most important limiting factor with respect to improving the satellite's response time is the commanding delay. While individual pointings and pre-loaded tiling patterns, like the 37-point tiling that covered the center of the *Fermi* GBM localization region, can be uploaded through the Tracking and Data Relay Satellite System (TDRSS) with a 25 minute delay at any time of the day, the more complex custom tiling to cover the GW error regions currently require a ground station pass. The time between these passes vary, but if one is unlucky and receives a trigger notification right after a pass, there may be a multiple hour wait before the next one. In the case of GW 170817, a pass began almost immediately after receiving the epochal GCN (LVC, GCN 21505), which is what allowed for the rapid 16 minute response time. The *Swift* team is currently in the late stages of testing a system, called PPTOO (Pre-Planned Target of Opportunity) that will effectively allow for any set of commands (including tiling campaigns for custom GW error regions) to be uploaded through the TDRSS system, which will ensure a 25 minute response time even in the worst case scenario of a multiple-hour wait until the next ground-station pass.

4.3. *Tiling Strategy*

Swift began its tiling campaign based on the 3-detector skymap at $\sim$ 18:00 UT, many hours before the sun set in Chile, meaning *Swift* was able to be on-target before most of the ground-based telescopes capable of observing the location of GW 170817. The host galaxy of AT 2017gfo, NGC 4993, was one of the most massive, and therefore most probable, galaxies in the field. Indeed NGC 4993 was in the first 10 galaxies observed by the ground observatories. Given the fact that *Swift* began observing hours earlier and spent at most 120 s per field, why did *Swift* not observe AT 2017gfo first?

The reason lies in the fact that *Swift* did not observe the galaxies in strict probability order. The tiling campaigns were constructed by optimizing the schedule on minimum slew distance as well as probability. This choice was made based on studies performed with the 2-detector skymaps from events in O1 and the beginning of O2. The localization regions from these events were quite large and were often bimodal and had regions that were spatially disconnected. Large sets of simulations were performed to determine the best way to tile these regions. It was found that the probability of finding a counterpart was actually increased if the list was weighted by slew distance in addition to probability.

Essentially, in the case of spatially disconnected probability regions the spacecraft can spend large amounts of time slewing back-and-forth across the sky, to the point that the observing-to-slew time ratio drops significantly enough such that the spacecraft can actually observe more cumulative probability in the same amount of time if one weights the list by slew distance.

However, GW170817 had a very small localization region, and therefore had *Swift* simply worked it's way down the galaxy list in strict probability order, it would have observed NGC 4993, and hence AT 2017gfo, at 18:09 UT on August 17, 5 hours before the discovery by ground telescopes. Given the knowledge that *Swift* can in fact find these counterparts and potentially provide extremely important observations at very early times, it is worthwhile to improve our response time to ensure that we do in fact succeed in this capacity in O3.

5. Conclusion

Swift's fast and wide response provided the only sensitive UV and X-ray transient search in the localization region, and increased the confidence that AT 2017gfo was in fact the counterpart to GW and sGRB 170817. Moreover, *Swift*'s early discovery of UV emission from AT 2017gfo and the absence of early X-rays provide independent constraints on the dynamics and orientation of the kilonova. *Swift* is a unique facility and the only satellite observatory capable of carrying out these rapid response custom tiling campaigns. *Swift* will continue its GW follow-up activities in O3 with significant improvements in many facets of its operations including commanding latency, scheduling, and tiling strategy. These operational improvements will also significantly benefit other multi-messenger and localization projects, and further ensure that *Swift* is able to maximally contribute to the new age of gravitational wave and multi-messenger astronomy.

References

Abbott, B. *et al.* 2017, *ApJ* **848**, 2.
Coulter, D. *et al.* 2017, Gamma Ray Coordinates Network Circular 21529.
D'Avanzo, P. *et al.* 2014, *MNRAS* **442**, 2342.
Evans, P. *et al.* 2012, *ApJS* **203**, 28.
Evans, P. *et al.* 2016, *MNRAS* **462**, 1591.
Evans, P. *et al.* 2017, *Science*, DOI: 10.1126/science.aap9580.
Gehrels, N. *et al.* 2004, *ApJ* **611**, 1005
Kasliwal, M. *et al.* 2017, *Science*, DOI: 10.1126/science.aap9455.
Smartt, S. *et al.* 2017, *Nature*, DOI: 10.1038/nature24303.
The LIGO Scientific Collaboration, the Virgo Collaboration 2017, Gamma Ray Coordinates Network Circular 21505.
The LIGO Scientific Collaboration, the Virgo Collaboration 2017, Gamma Ray Coordinates Network Circular 21513.
Tohuvavohu, A. 2017, *IOP: Proc. ACAT, submitted.*
van der Horst, A. J. *et al.* 2009, *ApJ* **699** 1087.

Gravitational Wave Astrophysics: Early Results from Gravitational Wave Searches and Electromagnetic Counterparts
Proceedings IAU Symposium No. 338, 2017
G. González & R. Hynes, eds.

doi:10.1017/S1743921318003599

DESGW Optical Follow-up of BBH LIGO-Virgo Events with DECam

Robert E. Butler[1], M. Soares-Santos[2], J. Annis[3], K. Herner[3]
for the Dark Energy Survey Gravitational Wave program and the DECam Community Group Collaboration

[1]Department of Astronomy, Indiana University, 727 E. Third Street, Bloomington, Indiana 47405, USA
[2]Martin A. Fisher School of Physics, Brandeis University, 415 South Street, Waltham, MA 02453, USA
[3]Fermi National Accelerator Laboratory, P.O. Box 500, Batavia, IL 60510, USA

Abstract. The DESGW program is a collaboration between members of the Dark Energy Survey, the wider astronomical community, and the LIGO-Virgo Collaboration to search for optical counterparts of gravitational wave events, such as those expected from binary neutron star mergers or neutron star-black hole mergers. While binary black hole (BBH) events are not expected to produce an electromagnetic (EM) signature, emission is certainly not impossible. The DESGW program has performed follow-up observations of four BBH events detected by LIGO in order to search for any possible EM counterpart. Failure to find such counterparts is still relevant in that it produces limits on optical emission from such events. This is a review of follow-up results from O1 BBH events and a discussion of the status of ongoing uniform re-analysis of all BBH events that DESGW has followed up to date.

Keywords. gravitational waves

1. Introduction

Are black hole mergers visible? Many observatories are engaged in following up LIGO events in an effort to localize the source and to begin study of the matter involved in the merger. Supermassive black holes often have accretion disks and jets—quasars are among the brightest objects in the universe. Stellar-mass black holes with accretion disks are visible in the x-ray, often remarkably so. Stellar-mass black holes without accretion disks may still be visible as they plow through the galactic interstellar medium, but they will be of magnitude ~ 22 in i-band at distance ~ 100pc (Heckler *et al.* 1996; Chisholm *et al.* 2003), rendering them invisible at the distances of the LIGO black holes. However there is the possibility of an unexpected optical counterpart, perhaps from the ubiquitous magnetic field induced jet.

Why DES and DECam? The combination of the collecting area of the 4m diameter Blanco Telescope at NOAO's Cerro Tololo International Observatory in Chile with the 3 deg^2 field-of-view of the Dark Energy Camera (DECam; Flaugher *et al.* 2015) gives us an unrivaled ability to cover large areas of sky to faint magnitudes; we are unique in the Southern hemisphere.

2. Methods and Results

Recap of O1. During O1, DESGW was able to follow up two events from LIGO (sky maps shown on the left in Figure 1). Among ground-based instruments, DESGW was the most complete in both depth and area covered. Our group and its partners have

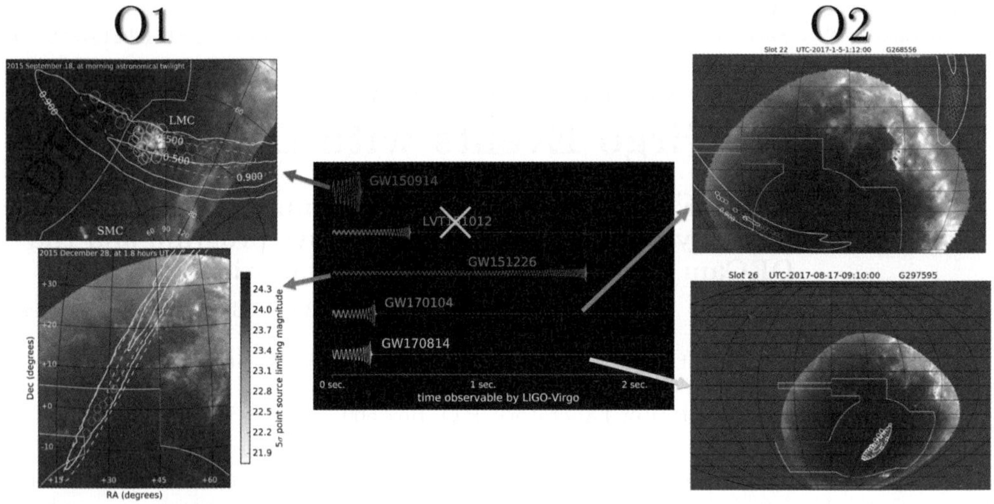

Figure 1. The middle panel (from LIGO's web gallery) shows the fitted chirp signals from each of the five gravitational wave events detected by LIGO so far. The two panels on the left show DESGW's observing plan for the two 2015 events, overlaid with the localization probability maps from LIGO. The two panels on the right show the same for the two 2017 events. Note that GW170814 was also observed by the Virgo detector, and therefore was localized much more precisely. Also note that LIGO did not send a trigger for the marginal LVT151012 event.

published on these events (GW150914: Soares-Santos *et al.* 2016, Annis *et al.* 2016; GW151226: Cowperthwaite *et al.* 2016). While the total probability coverage (P_{total} in Table 1) was quite small for each event in O1, the exercise and analysis were critical for future endeavors in DECam optical follow-up.

Improvements for O2. Between the first and second LIGO observing runs (and continuing throughout O2), the DESGW team made marked improvements in both our difference imaging (`diffimg`; Kessler *et al.* 2015) and post-processing pipelines. Both pipelines are described in further detail in Herner *et al.* (2017).

Difference Imaging. There were two main improvements in the `diffimg` pipeline over the past year: (1) the ability to run nightly, or "on the fly"; and (2) incorporation of the ability to use templates from *after* a given event to subtract from that event's search images. We can now image the area at a much later time and use these as templates. We have now done this using the BLISS Survey (NOAO propID 2017A-0260) for the first three BBH events with good success; we effectively doubled our ε_{camera} (see Table 1) for GW150914 with this strategy.

Post-processing. We also completely reworked the post-processing pipeline (`postproc`). The `postproc` code now creates a number of plots useful for assessing detections and `diffimg` efficiency. Two of these plots are shown in Figures 2 and 3. The code also has a new step which checks the outputs of `diffimg`, allowing the user to quickly and easily diagnose problems in the pipeline.

Update on O2. During O2, we followed up two LIGO triggers (as shown in Figure 1). GW170104 was not well placed for observations from the Blanco telescope, but we managed 14% of the spatial localization area. Analysis of that event is ongoing. GW170814 was better for us, since the new smaller probability contours from LIGO+Virgo lay directly in the middle of our survey footprint. This is obviously helpful observationally, but high-quality template images are also necessary for effective difference imaging. The DES

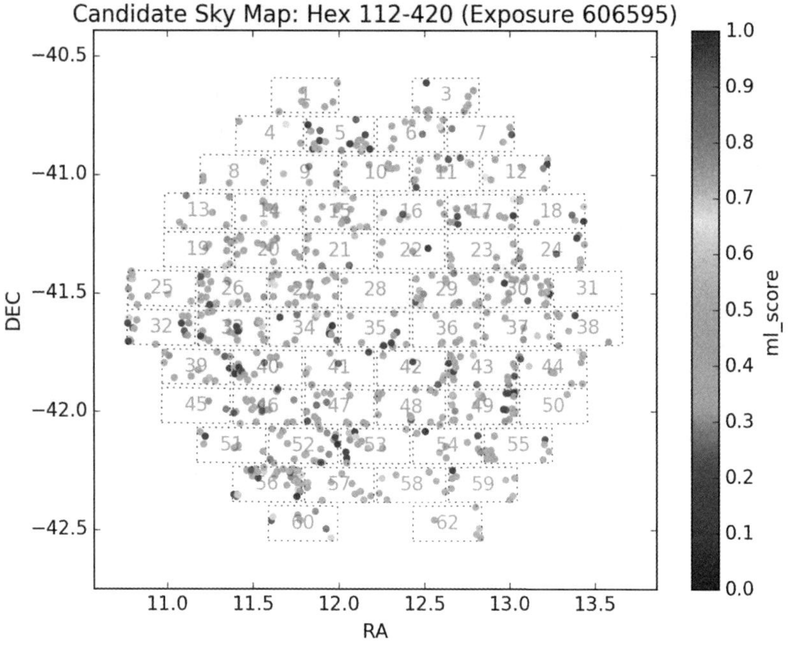

Figure 2. Example hex map output by the postprocessing pipeline. It shows all candidates within a certain exposure, with their associated machine learning scores (these are an assessment of how good the detection was).

Table 1. For each event, three statistics are given (defined in Annis & Soares-Santos 2016): (1) $\Sigma_{spatial}$, the summed probability inside the LIGO spatial localization map covered by the bounding box of images taken; (2) ε_{camera}, the fraction of the space within the bounding box which is covered with live CCDs, and (3) ε_{sky}, the fraction of the area imaged that a source would have been visible in (dominated by limiting magnitude). The multiplication of these three gives the probability that the imaging element was able to detect the source: P_{total}.

Event	$\Sigma_{spatial}$	ε_{camera}	ε_{sky}	P_{total}
GW150914	0.11	$0.8 \cdot 0.34$	0.8	0.024
GW151226	0.02	0.8	0.8	0.013
GW170104	0.14	0.96	0.8*	0.11*
GW170814	0.84	0.96	0.8*	0.65*
Total				***.790****

*Preliminary. We have not yet made a definite calculation of ε_{sky} for the most recent two events, but it is reasonable to expect a value of 0.8. Using that assumption, we get the total probabilities given in the rightmost column.

footprint comprises the majority of DECam images taken to date. As shown in Table 1, we were able to cover 84% of the final map (it was shifted slightly from the original).

It should also be noted that a third O2 BBH event was announced recently (after the conference), labeled GW170608. We did not follow up this event because the bulk of the probability was much too far north to be reached.

3. Conclusion

We have followed up four BBH events, and to date no viable candidates have been found in the data. There is still analysis to be done. For our purposes, LIGO events are

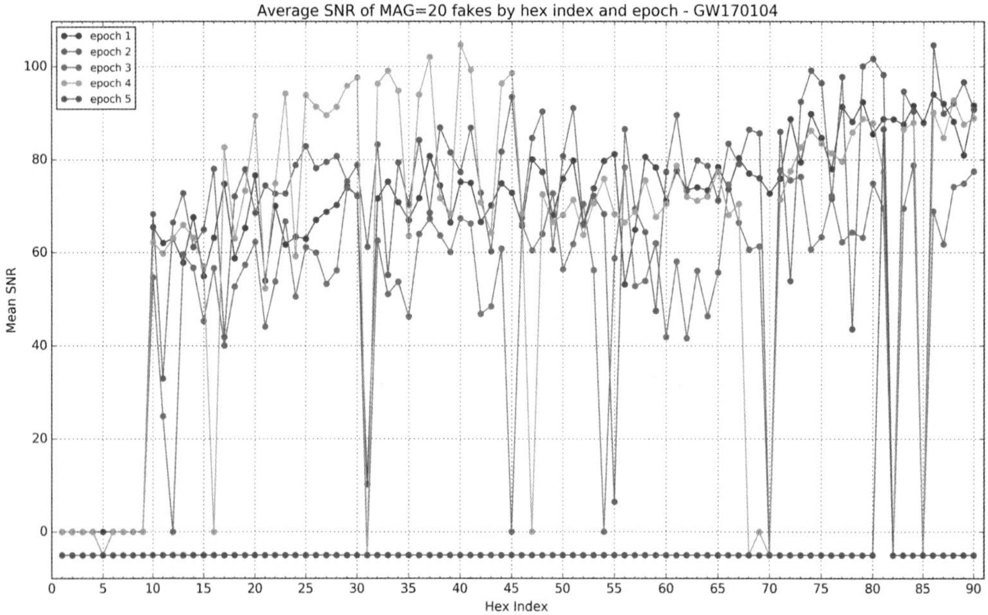

Figure 3. Signal-to-noise ratio (SNR) of artificially implanted signals (fakes) of constant magnitude for the LIGO event GW170104. These artificial transients are arranged on a uniform grid across the sky. The horizontal axis is labeled with indices representing the 90 different pointings on the sky. The vertical axis shows the average SNR of all fakes within a certain pointing, separated into five epochs. For most pointings, four epochs were observed. On this plot, mean SNR values of 0 and -5 are assigned to exposures which have not been processed or taken.

cumulative. If one wants to place a 95% confidence limit at a given magnitude, $P_{total} \geqslant 3$ is necessary. We are well on our way to this goal. With the increased detection rate expected in the next LIGO-Virgo run, it is reasonable to expect this milestone can be reached within that run.

References

Annis, J. & Soares-Santos, M. 2016, PoS (ICHEP2016) 1217

Annis, J., Soares-Santos, M., Berger, E. *et al.* 2016, *ApJ* (Letters) 823, L34

Chisholm, J. R., Dodelson, S., & Kolb, E. W. 2003, *ApJ*, 596, 437

Cowperthwaite, P. S., Berger, E., Soares-Santos, M., Annis, J. *et al.* 2016, *ApJ* (Letters) 826, L29

Flaugher, B., Diehl, H. T., Honscheid, K. *et al.* 2015, *AJ*, 150, 150

Heckler, A. F. & Kolb, E. W. 1996, *ApJ* (Letters) 472, L85

Herner, K. *et al.* 2017, *J. Phys.: Conf. Ser.* 898, 032050

Kessler, R., Marriner, J. *et al.* 2015, *AJ* 150, 172

Soares-Santos, M., Kessler, R., Berger, E., Annis, J. *et al.* 2016, *ApJ* (Letters) 823, L33

Gravitational Wave Astrophysics: Early Results from Gravitational Wave Searches and Electromagnetic Counterparts
Proceedings IAU Symposium No. 338, 2017
G. González & R. Hynes, eds.

doi:10.1017/S1743921318003794

Cosmology with Gravitational Waves in DES and LSST

Kenneth Herner[1], Marcelle Soares-Santos[2,1], James Annis[1]
on behalf of the Dark Energy Survey Gravitational Wave program and the DECam Community Group Collaboration

[1]Fermi National Accelerator Laboratory, Batavia, IL 60510, USA
email: kherner@fnal.gov

[2]Dept. of Physics, Brandeis University, Waltham, MA 02453, USA

Abstract. Motivated by the prospect of the wealth of data arising from the inauguration of the era of gravitational wave detection by ground-based interferometers the DES collaboration, in partnership with members of the LIGO collaboration and members of the astronomical community at large, have established a research program to search for their optical counterparts and to explore their use as cosmological probes. In this talk we present the status of our program and discuss prospects for establishing this new probe as part of the portfolio of the Dark Energy research program in the future, in particular for the next generation survey, LSST.

Keywords. (cosmology:) cosmological parameters, gravitational waves, stars: neutron, image processing

1. Introduction

One of the numerous breakthroughs that come from the dawn of gravitational wave (GW) astronomy is the ability to measure cosmological parameters. Perhaps the most natural parameter to measure is the Hubble Constant H_0, the expansion rate of the local universe. In turn the value of H_0 has consequences for other cosmological parameters such as the dark energy equation of state.

Merging compact binaries such as binary black hole (BBH) mergers, binary neutron star (BNS) mergers, and neutron star-black hole (NSBH) mergers can be used as so-called "standard sirens", to measure H_0 (Schutz 1986, Holz and Hughes 2005). Compact binaries can be broadly classified into two categories, "bright" and "dark" standard sirens, depending on whether they are expected to have an electromagnetic signature.

1.1. *Bright standard siren cosmology*

BNS and NS-BH events are known as "bright" standard sirens since there is an electromagnetic counterpart to the GW signal. Figure 1 illustrates how one can combine information from the GW waveform and from observations of the EM counterpart to measure the Hubble constant (Nissanke 2013). By identifying this counterpart and its host galaxy, if applicable, one can determine a redshift, and thus a recessional velocity as $v = cz$ at low redshift. After correcting for the peculiar velocity of the host relative to the background Hubble flow, one is left with a Hubble velocity v_H. This in turn leads to a determination of the Hubble constant via the relation $v_H = H_0 d$, where d is the distance to the source as measured from the GW signal.

The LIGO/Virgo Collaboration (LVC) has established partnerships with several collaborations around the world that perform observations across the EM spectrum to search for counterparts to GW event. Shortly after their initial analysis of a GW candidate, the

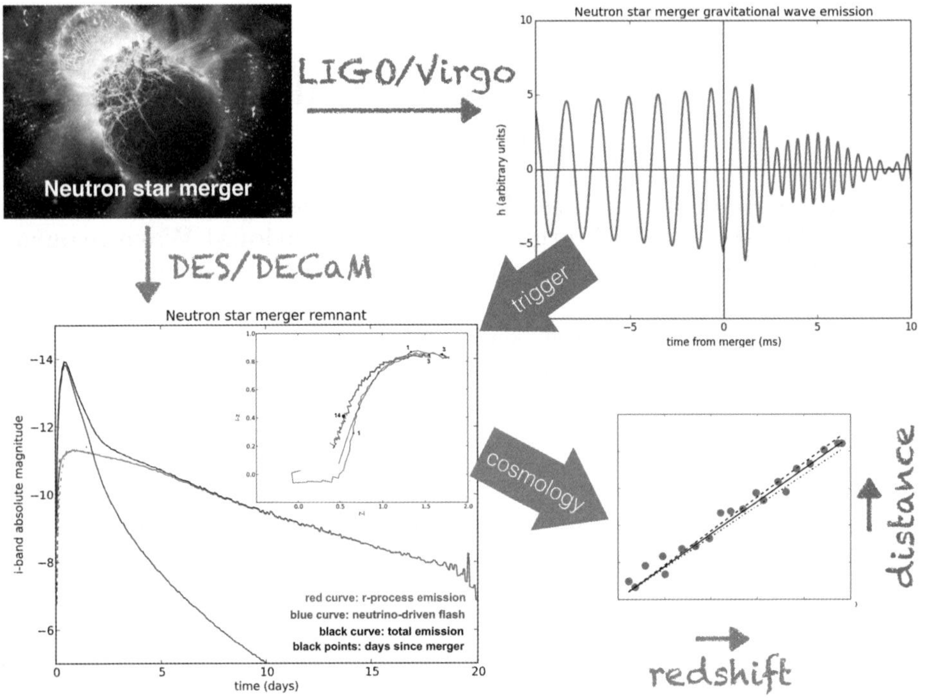

Figure 1. Diagram of the process to measure H_0 with bright standard sirens (BNS and NS-BH events). One can combine absolute distance measurement from the GW waveform and the redshift of the optical counterpart (or its host galaxy) and solve for H_0 using the relation $v_H = H_0 d$, with $v_H = cz$ at low redshift.

LVC shares information with these partners, including estimated distance to the source, most likely type (e.g. BBH, BNS), and a map showing the probability of the source to be located at each point on the sky. The partner collaborations use this information to plan their follow-up campaigns.

1.2. *Dark standard siren cosmology*

BBH events can also form the basis of a H_0 measurement. In this case one can create a spatial map of BBH events using the localization and distance from LIGO. One can also make a map of galaxies and their redshift, as shown in Figure 2. By assuming a value for H_0, one can calculate a cross-correlation between the two maps. Varying the assumed value of H_0, one can build a cross-correlation distribution with the true value of H_0 being where the distribution is at its maximum.

2. Overview of DESGW program

One of the LVC EM followup partner groups is the Dark Energy Survey Gravitational Wave group (DESGW). The DESGW group consists of members of the Dark Energy Survey (DES) and other members of the astronomical community. DES built and is one of the primary users of DECam (Flaugher *et al.* 2015), at the 4m Blanco Telescope at the Cerro Tololo InterAmerican Observatory in Chile. DECam is well-suited to the EM followup task due to its red-sensitive CCDs and its large aperture (≈ 3 deg^2 FOV). Once a trigger is received, our team uses a dedicated strategy code to decide whether or not to observe, how much telescope time to invest, and what sequence of observations

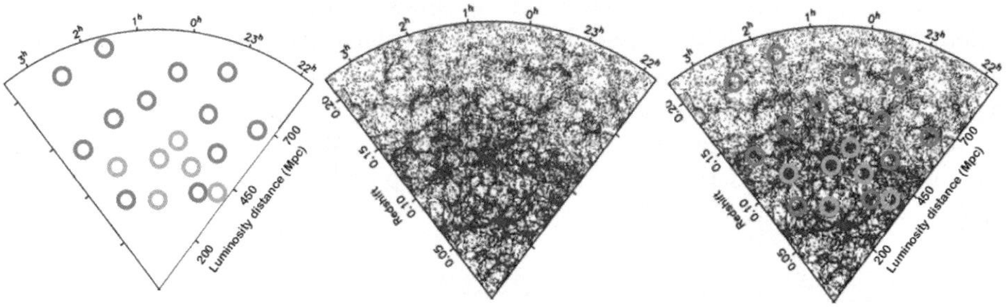

Figure 2. Measuring H_0 with dark standard sirens (BBH events). By comparing a map of BBH events (left panel; distances from LIGO measurements) with a map of galaxies (center panel; redshifts from catalogs), one can determine the value of H_0 by varying it and finding the value that maximizes the cross-correlation.

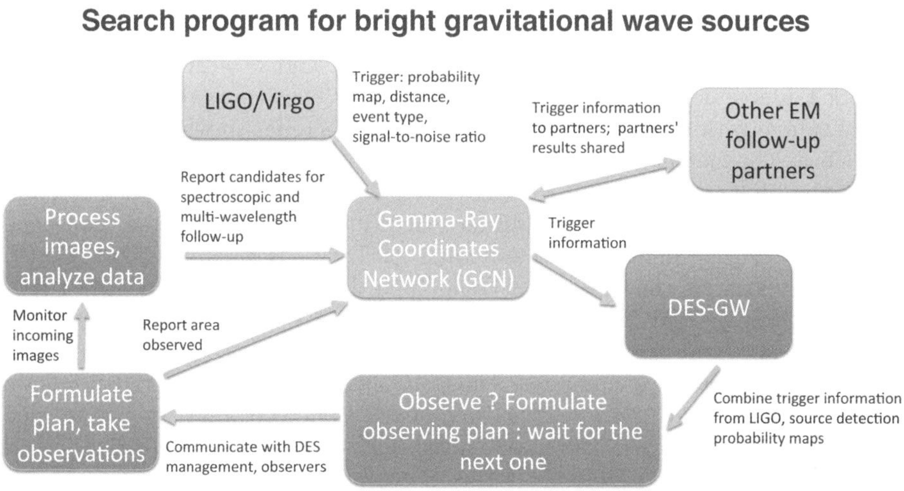

Figure 3. Overview of DESGW follow-up program. Communication between LIGO and EM partners is via GCN circulars. DESGW aims to complete image processing and candidate identification within 24 hours of data-taking.

to execute. If the decision is to observe, we interrupt DES observations to perform the search. Figure 3 illustrates the entire process from LIGO trigger, sent over the Gamma-ray Coordinates Network (GCN), to observations and candidate reporting, also via GCN circulars. The initial process is the same for BBH, BNS, and NS-BH events.

See Herner *et al.* (2017) for further details of the observing plan construction and source models. In particular, we note that due in part to DECam's relatively wide field of view, we do not follow a strategy of targeting only galaxies consistent with the LIGO distance estimate. Instead we follow a "blind" search strategy, considering only the probability of detection based on expected source brightness and observing conditions.

To identify counterpart candidates we use the `diffimg` pipeline, described in Kessler *et al.* (2015). The pipeline takes search images and subtracts one or more "template" images taken of the same region of sky. Typically these images are earlier in time than the search images, but it is possible to use images taken after the search image (long enough for any expected candidate objects to have faded below detection threshold) as templates if needed. We use all available DES images passing data quality requirements and $\geqslant$ 30s in length, as well as publicly available DECam exposures of at least 30s taken on non-DES time.

Since the light from BNS events fades on the timescale of days, it is critical to rapidly complete image processing and candidate identification so that we can report any candidates in a GCN circular to alert other groups that may perform detailed spectroscopic followup. In particular we aim to complete analysis of each night's images in no more than 24 hours, with less than 12 hours being preferred. We have access to considerable grid computing resources to make such timescales possible, and we have succeeded in hitting our targets in O2.

3. Results from LIGO trigger followup campaigns

3.1. *Followup of BBH events*

As of the end of August 2017, DESGW performed four BBH optical counterpart searches, two in O1 and two in O2. Figure 4, upper left panel, shows the sky map for the GW150914 (Abbott *et al.* 2016a) followup campaign. The main counterpart search, described in Soares-Santos *et al.* (2016), covered approximately 40 $\deg^2$ with 90s exposures in i and z. We are able to reach a depth of $i = 22.1$ and $z = 21.2$ in the subtracted images, for a source detection completeness of $> 80\%$. That translates into depth of 200 Mpc for a source with absolute magnitude, $M = -14$, similar to what one might expect for a kilonova. The merger emission model of Figure 1 assumes a search depth of 30Mpc. Note that the LIGO sensitivity in 2015–2016 was 40–80Mpc for binary neutron stars.

Before it was known that GW150914 was a BBH merger, we considered that respected the spatial localization maps intersection with the Large Magellanic Cloud (LMC) and developed a hypothesis that the event could be a dark core collapse in the LMC. We imaged 34 $\deg^2$ of the LMC region (Figure 4, upper right panel) with 5s exposures and searched for disappearance of previously cataloged red giant stars, the most likely to suffer a core collapse. All were accounted for. The LMC analysis Annis *et al.* (2016) allowed us to develop a template for future searches in case LIGO detects a nearby burst event and complements the main search. Our two analyses formed the most comprehensive optical search for signatures of GW150914 in terms of volume searched and methodology.

For the second LIGO event (GW151226; Abbott *et al.* 2016b), we performed only the main analysis (Figure 4, lower left panel) and covered a much smaller area because the region of interest was primarily in the north (Cowperthwaite *et al.* (2016)). That region also lacked pre-existing templates so we adapted our program to enable difference imaging with templates taken weeks after the trigger, when sources are expected to have faded. The analysis initially found four candidates; all were rejected as background upon further inspection. The longer timescale due to lack of full template coverage further motivates the pursuit of all sky-coverage with DECam. For this event a large fraction of the image processing was performed on local campus resources at Harvard and compared with the grid runs.

During the second LIGO observing season, O2, DESGW followed up GW170104 (LIGO Scientific Collaboration and Virgo Collaboration 2017a) and GW170814 (LIGO Scientific Collaboration and Virgo Collaboration 2017b). The follow-up search areas covered 13.6% and 84% of the LIGO 90% probability regions, respectively. We are analyzing both events to search for any possible electromagnetic counterparts. We are also performing a combined analysis of all BBH events we have studied to date; the chief goal is to set upper limits on optical emission from BBH events.

3.2. *Followup of BNS events*

The first GW signal from a BNS event came on 17 August 2017 in the form of GW170817 (LIGO Scientific Collaboration and Virgo Collaboration 2017c), and triggered a massive

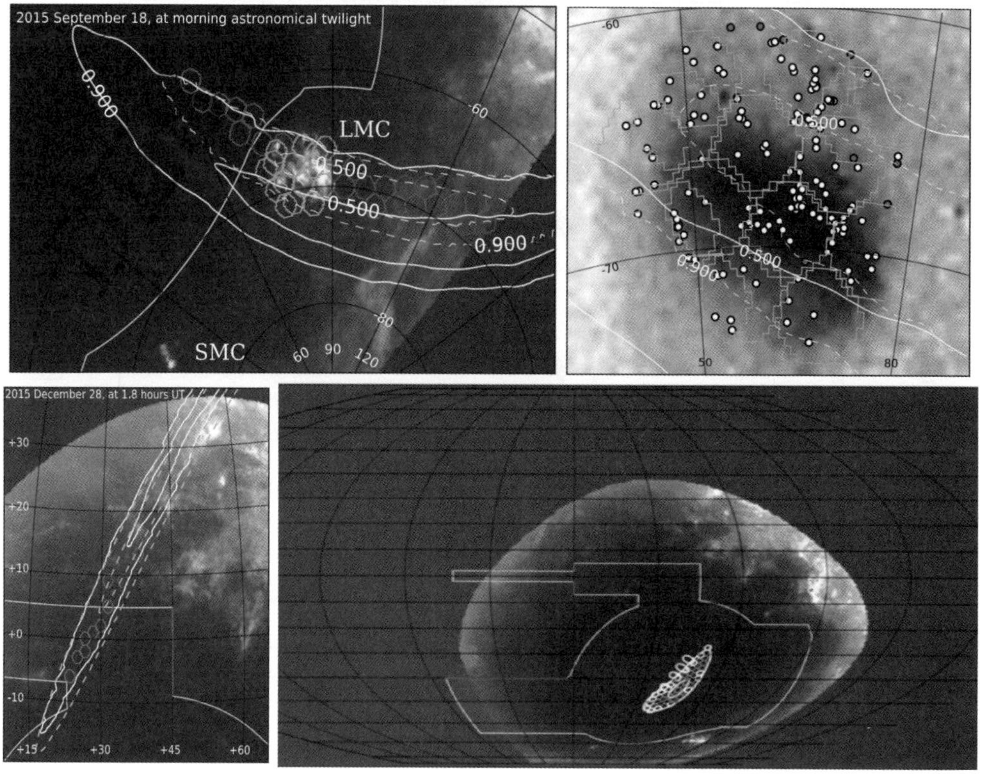

Figure 4. Sky maps for main followup of GW150914 (upper left), the disappearing star search in the LMC after GW150914 (upper right), the followup of GW151226 (lower left) and GW170814 (lower right). The white contours represent the 50% and 90% probability regions of the original (dashed) and updated (solid) LIGO probability maps. The yellow lines represent the fiducial DES observing region. The hexagons represent the locations of the telescope pointings during the search. In the upper left and both lower plots, the background colors indicate the DES limiting magnitude for observation, and in the upper right plot the background colors indicate stellar density.

EM follow-up campaign around the world Abbott *et al.* 2017. DESGW was one of several groups to independently identify the optical counterpart, and was able to exclude all other potential candidates within a large fraction (> 80%) of the LIGO probability region (Soares-Santos *et al.* 2017. We also performed detailed follow-up observations for nearly two week after the merger, and compare the data to several possible kilonova models in Cowperthwaite *et al.* (2017).

Our blind search strategy also enables the use of the redshift of the associated galaxy (NGC 4993) to calculate a value of H_0 in conjunction with the source distance estimate from the GW signal in LIGO Scientific Collaboration and Virgo Collaboration (2017d). The calculated value is $70.0^{+12.0}_{-8.0}$ km/s/Mpc, consistent with results from supernovae (Reiss *et al.* 2016) and the CMB (Planck 2015), while also being independent of the cosmic distance ladder.

4. Future directions

The third LIGO observing run is scheduled to begin in the summer of 2018. In the interim the DESGW group is working on a number of enhancements to its own

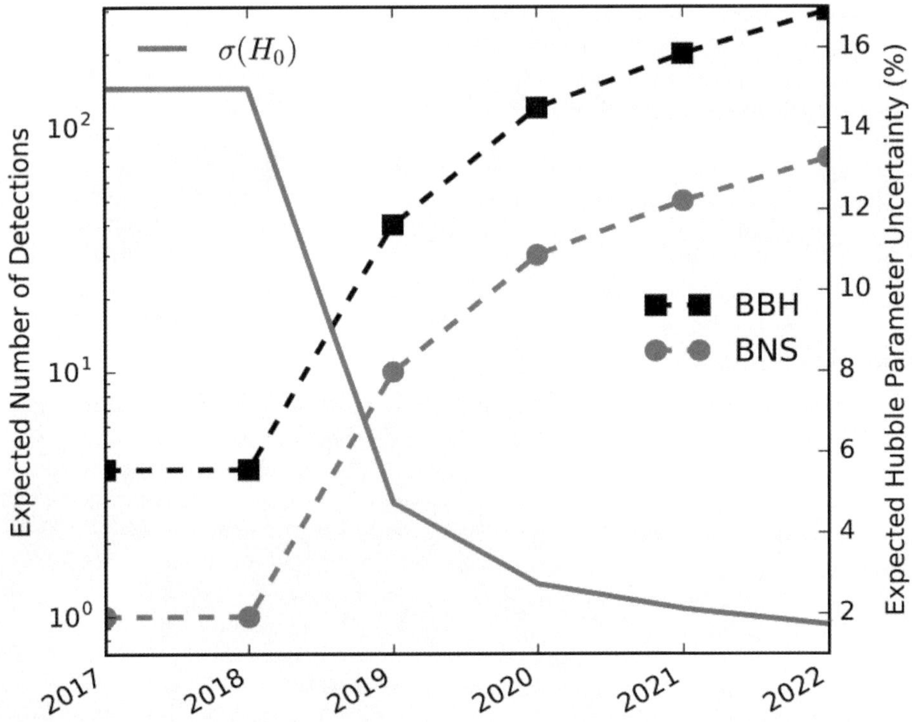

Figure 5. Expected number of BBH events (dark dotted line) and BNS events (light dotted line) detected by LIGO as a function of time. The solid line represents the expected uncertainty on H_0, assuming improvements in the statistical uncertainty due to the larger number of events.

program. These include gathering template images in regions where there is no existing DECam coverage, scaling up the overall data transfer rates, and building up a catalog of galaxy peculiar velocity measurements. This last item is particularly important for improving the accuracy of future H_0 measurements, since one must account for the peculiar velocity of a counterpart's host galaxy in order to obtain an accurate Hubble velocity.

The BNS detection rate will rise as LIGO's sensitivity increases. Starting from the present uncertainty of $\approx 15\%$ on H_0 measured from GW170817, and extrapolating from Figure 2 of the DAWN II report (LSC 2016) and assuming only improvements in uncertainty due to statistics, we can estimate the expected precision of GW-based H_0 measurements from KNe over time, shown here in Figure 5. We find that by 2019, we could expect the uncertainty to be at approximately 5%, and by the start of the LSST era in 2022, a precision of $< 2\%$ should be possible with roughly 60 events.

Another possibility involves searches for untriggered kilonovae. The authors of Scolnic *et al.* (2017) have estimated how many KNe one might expect to identify without a LIGO trigger in several survey programs, including DES, WFIRST, and LSST. They find that relatively few kilonovae in completed and ongoing survey could be seen without a GW trigger, but several dozen could be seen with future surveys such as WFIRST and LSST. Indeed, WFIRST in particular should allow for kilonova detections out to $z \approx 0.8$, past the sensitivity of current GW detectors.

5. Conclusions

A source with the same absolute magnitude as GW170817, $M_i = -15.7$, would be bright enough for the DESGW program to detect it out to 425 Mpc. This detection has ushered in the era of multi-messenger astronomy, and enables independent measurements of cosmological parameters such as the Hubble constant. As GW detectors improve their sensitivity and additional join GW-EM observations occur, percent-level measurements of cosmological parameters should be possible at the dawn of the LSST era. Future survey projects, especially when combined with next-generation GW detectors, should provide an unparalleled window into the cosmos.

Acknowledgments

This research uses services or data provided by the NOAO Science Archive. NOAO is operated by the Association of Universities for Research in Astronomy (AURA), Inc. under a cooperative agreement with the National Science Foundation. This manuscript has been authored by Fermi Research Alliance, LLC under Contract No. DE-AC02-07CH11359 with the U.S. Department of Energy, Office of Science, Office of High Energy Physics.

References

Abbott, B. P. *et al.*, 2016, *Phys. Rev. Lett.*, 116, 061102
Abbott, B. P. *et al.*, 2016, *Phys. Rev. Lett.*, 116, 241103
Abbott, B. P. *et al.*, 2017, *Phys. Rev. Lett.*, 118, 221101
Abbott, B. P. *et al.*, 2017, *Phys. Rev. Lett.*, 119, 141101
Abbott, B. P. *et al.*, 2017, *Phys. Rev. Lett.*, 119, 161101
LIGO Scientific Collaboration and Virgo Collaboration *et al.*, 2017, *Nature*, 551, 85-88
Abbott, B. P. *et al.* 2017, *ApJ* (Letters), 848, L12
Annis, J. *et al.* 2016, *ApJ* (Letters), 823, L34
Cowperthwaite P. S., *et al.* 2016, *ApJ* (Letters), 826, L29
Cowperthwaite, P. S. *et al.* 2017, *ApJ* (Letters), 848, L17
Flaugher, B. *et al.* 2015, *AJ*, 150, 150
Herner, K. *et al.* 2017, *J. Phys.: Conf. Ser.*, 898, 032050
Holz, D. E. & Hughes, S. A. 2005, *ApJ*, 629, 15
Kessler R., Marriner J., Childress M. *et al.* 2015, *AJ*, 150, 172
LIGO Scientific Collaboration 2016, LIGO-P1600350
Nissanke, S. *et al.* 2013, arXiv:1307.2638
Planck Collaboration 2016, *A&A*, 594, A13
Riess, A. G. *et al.* 2016, *ApJ*, 826, 56
Soares-Santos M., *et al.* 2016, *ApJ* (Letters), 823, L33
Soares-Santos, M. *et al.* 2017, *ApJ* (Letters), 848, L16
Schutz, B. F. 1986, *Nature*, 323, 310
Scolnic, D. *et al.* 2017, *ApJ* (Letters), in press.

Gravitational Wave Astrophysics: Early Results from Gravitational Wave Searches and Electromagnetic Counterparts
Proceedings IAU Symposium No. 338, 2017
G. González & R. Hynes, eds.

doi:10.1017/S1743921318000935

Observations of GW170817 by DESGW and the DECam GW-EM Collaboration

James Annis

Fermi National Accelerator Laboratory
Batavia, Illinois, USA
email: annis@fnal.gov

Abstract. On August 17, 2017 LIGO/Virgo detected a binary neutron star via gravitational waves. We observed 70 sq-degrees in the LIGO/Virgo spatial localization with the DECam on the 4m Blanco telescope covering 80% of the final map. Our group independently discovered an optical counterpart in NGC 4993. We searched our entire imaged region: the object in NGC 4993 was the only viable candidate. Our observations of NGC4993 show complicated morphology but simple star formation history. Our x-ray and radio observations indicate an off-axis jet as afterglow. Our high-cadence optical and infrared spectra show a source that must be described by at least two components, one of which is dominated by the r-process nucleosynthesis elements characteristic of a kilonova. Our modeling of the light curve demonstrates such a model in which 0.05 $M_\odot$ of material is ejected from the system. Finally, we discuss the first standard siren measurement of H_0.

Keywords. gravitational waves, (cosmology:) cosmological parameters.

1. Introduction

The DESGW and the DECam GW-EM Collaboration use of the Dark Energy Camera (DECam; Flaugher *et al.* 2015), a 3.1 deg^2 wide-field imager on the prime focus of the Blanco 4m telescope at the Cerro Tololo Inter-American Observatory, to perform wide area searches for the counterparts of LIGO/Virgo gravitational wave (GW) sources.

DESGW performs the search and discovery. LIGO/Virgo has published two triggered binary black hole (BBH) mergers from their observing run O1: GW150914 and GW151226, and three from O2: GW170104, GW170608, and GW170814 (respectively Abbott *et al.* 2016, Abbott *et al.* 2017a, Abbott *et al.* 2017c, Abbott *et al.* 2017d, and Abbott *et al.* 2017e). We have published DECam based optical searches on GW150914 (Soares-Santos *et al.* 2016, Annis *et al.* 2016), GW151226 (Cowperthwaite *et al.* 2016), and are pursuing analyses on GW170104 and GW170814. One of aims of DESGW is to establish standard sirens (Schutz 1986, Holz & Hughes 2005) as a cosmological tool. The poster by Butler *et al.*, and the talk by Herner *et al.*, these proceedings, have more details on our BBH observations and DESGW.

The strength of the DECam GW-EM Collaboration is the wide array of telescopes that can be brought to bear on a GW event. For GW170817 we obtained data across the electromagentic spectrum: in radio the VLA and ALMA; in the optical and NIR the Hubble Space Telescope 2.4m, the Blanco 4m, the SOAR 4m, the Gemini 8m, the Magellan 8m; in the x-ray the Chandra spacecraft. Our collaboration has long experience following up short gamma ray burst events, and the expected optical counterparts of neutron star merger events (see, e.g., Metzger & Berger 2012, Berger 2014).

2. Search and Discovery

On August 17, 2017 the LIGO/Virgo GW detectors recorded the loudest event yet, a long, high frequency waveform. The interpretation was of a binary neutron star merger of 1.5 + 1.3 $M_{\odot}$, resulting in a merger remnant of 2.7 $M_{\odot}$, heard from a distance of 40 Mpc. As the three detector network was operating, the spatial localization area was small (Abbott *et al.* 2017f).

An optical and infrared counterpart to the NS event GW170817 was independently detected by several teams within the next 12 hr (Abbott *et al.* 2017g), including ours (Allam *et al.* 2017), at a location deep inside the bright galaxy NGC 4993. In our collaboration, the source was found via a visual inspection of nearby galaxies in our raw data stream by Ryan Chornock. The discovery by DESGW initiated a follow-up program by our collaboration that lasted until the counterpart was lost behind the sun two weeks later. (Soares-Santos *et al.* 2017).

We performed a search for all possible counterparts in our data. Our search is designed to find kilonova as predicted in, e.g., Barnes & Kasen 2013. We imaged 70 deg^2 in i and z bands, in two tilings of 18 hexes at 90sec in each, deep enough to find kilonova out to > 100 Mpc despite the uncertainties in the predicted luminosities of kilonova. The search area was only 40% covered by pre-existing DECam images and we acquired the rest August 31 to September 2. We expect any kilonova-like transients to have faded below detection thresholds by that time, > 2 weeks after the trigger. The camera fill factor is 80%, leaving 4% area loss after two tilings. Weather on the nights of template observations was partially clouded and caused a 3% loss. Processing failures accounted for less than 1% loss. The final area included in the search was 65 square degrees. Our difference imaging results in 1500 transient candidates with magnitudes between 15.5 and 20.5, where a candidate means on object detected on at least two search images. The magnitude cutoff was limited by the depth of the template images: i = 21.2 and z = 20.5, which can be compared with the depth of the search images: i = 22.0 and z = 21.3. This list of transients was cleaned by a set of cuts designed to reject moving objects, background artifacts and long-lived transients:

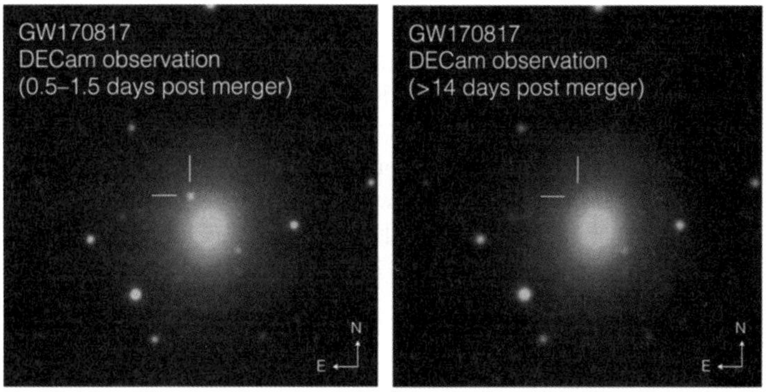

Figure 1. The DESGW discovery image of the NGC4993 kilonova (Soares-Santos *et al.* 2017).

(*a*) The candidate must have at least one detection in i and one detection in z band.

(*b*) The candidate must pass our machine learning based junk rejection threshold. This rejects non-astrophysical artifacts such as dipoles. The efficiency of this criterion as measured from point sources injected into our images is $\sim 90 - 100\%$ for both i and z $\sim 17 - 22$ mag.

(*c*) The candidate must have faded significantly between the search (first) and template (last) observations. Specifically, we require the change in the candidate flux to be greater, at 3σ level, than the flux in a circular aperture of radius = 5 pixels on the template image. This criterion eliminates slow-evolving transients (e.g., supernovae).

The initial list of 1500 candidates are cut down to 252 by the first cut, down to 81 candidates by the second cut, and down to one candidate by the third cut. This candidate is the optical transient in NGC4993. We therefore conclude that the counterpart to GW170817 is the new object discovered in NGC 4993. (Soares-Santos *et al.* 2017).

The first DECam images of the optical counterpart to the GW170817 at 11.40 hours post-merger showed an object with $i = 17.30$ mag and $(i - z) = 0.13$, bright and nearly flat spectrum, with an absolute magnitude of $M_i = 15.7$. For the first few days the optical data are consistent with a blackbody, and for the first few days the data in each bandpass falls $m \propto t^{-n}$, where n ranges from 2 in the u-band to 0.4 in the Y-band. The object fades much faster in the blue than it does in the red. Our observations on day 0.6 and on day 1.5 show that the peak of the light curve of this event was at ≈ 1 day. At $M_i = 15.7$, the optical transient is bright enough for us to detect it out to 425 Mpc. (Soares-Santos *et al.* 2017).

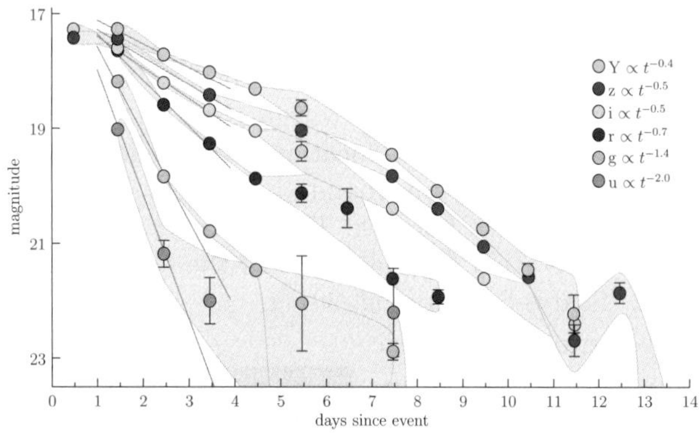

Figure 2. DECam observations of the counterpart in NGC 4993 for the two weeks after the merger (Soares-Santos *et al.* 2017).

3. NGC 4993: The Host Galaxy

We used DECam (Palmese *et al.* 2017) and Hubble Space Telescope data (Blanchard *et al.* 2017), along with archival data, to study the host galaxy of GW170817. The HST data in particular rule out a point source at the location of the counterpart brighter than $M_V = -7$ as recently as 4 months prior to the merger; the eliminates supergiants as a progenitor and likely globular clusters as the site of the merger as half the globular cluster population would be brighter than this.

The NGC 4993 is an elliptical galaxy that in i-band is well described by a $n \approx 4$ Sersic profile. The offset of the counterpart from the nucleus of NGC 4993 is 2.1 kpc, inside the galaxy's half-light radius. From modeling of the UV to MIR SED we infer the star formation history, and find an exponentially declining SFH, with a median stellar population age of $11.2^{+0.7}_{-1.4}$ Gyr. The present-day SFR is low, $\approx 0.01 M_{\odot}\ yr^{-1}$. Assuming that the star formation history is the probability distribution function for the binary birth age, the median merger timescale

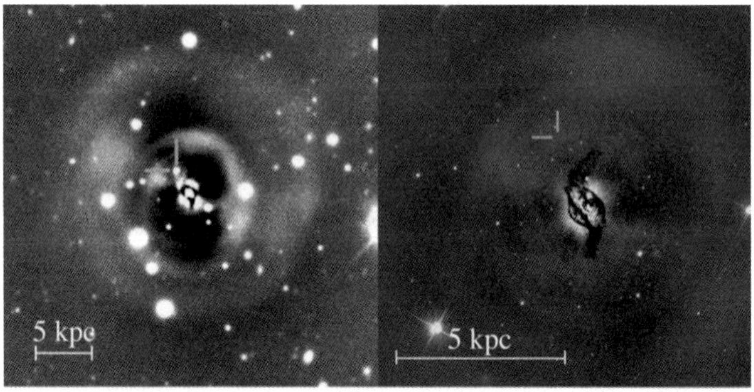

Figure 3. Shell, left, and central dust structure, right of NGC 4993. The counterpart location is marked (Palmese *et al.* 2017).

is $11.2^{+0.7}_{-1.4}$ Gyr with a 90% probability the BNS system formed between 6.8 – 13.6 Gyr ago. Assuming a circular orbit and equal neutron star masses of 1.4 $M_\odot$, this corresponds to an median initial separation of 4.5 $R_\odot$. The resulting initial separation and eccentricity constraints are consistent with current galactic binary NS systems. Given the long merger timescale and small projected offset we conclude that the binary system experienced at most a modest natal kick, with an upper limit of 200 km/sec. (Blanchard *et al.* 2017).

The DECam data shows NGC 4993 has clear and significant shell structure. Shell structure in indicative of a minor merger, in which the smaller galaxy has $< 10\%$ of the mass of the larger galaxy. The HST data shows complex dust lanes in the galactic center. A spectrum from 6dF shows narrow H_α, H_β, OIII, & NII lines whose line ratios indicate the presence of a AGN, as does the x-ray and radio luminosites of the center. These indicate that the smaller galaxy was dust and gas rich in comparison with NGC 4993, and the merger transported these to the center while the stars formed the shells. The position of the counterpart on or near one of the shells; there is no star formation occurring there. The merger likely happened 28 Myr ago. These features are uncommon in ellipticals, and it is interesting that the NS merger happened in such a galaxy. (Palmese *et al.* 2017).

4. The Counterpart as a Short GRB Afterglow

Two seconds after the GW merger event, two spacecraft carrying gamma ray detectors, Fermi's GBM and INTEGRAL's SPI-ACS, detected a short gamma ray burst (sGRB), and thus the collaborations operating them confirmed binary neutron star mergers as a progenitor of sGRBs. (Abbott *et al.* 2017h).

We obtained x-ray data on the counterpart using the Chandra spacecraft Margutti *et al.* 2017 and radio data from the Very Large Array and ALMA Alexander *et al.* 2017 to make observations of the sGRB afterglow.

The sGRB afterglow is broad brand synchrotron emission, and observations from radio to x-ray can constrain the spectral energy distribution and hence the physical properties of the event. We obtained the first x-ray detection from GW170817 using the Chandra spacecraft (Margutti *et al.* 2017), establishing a rising x-ray light curve. We also obtained extensive radio observations of GW170817 at centimeter and millimeter wavelengths using the VLA and ALMA, including the earliest observations taken in these bands. (Alexander *et al.* 2017). Combined as a function of time, three things stand out: 1) An on-axis afterglow emission similar to that typically observed in sGRBs ($E_{k,iso} \sim 10^{51}$erg) is clearly ruled out; 2) A late on-axis or isotropic afterglow onset due to the deceleration of a mildly relativistic outflow is consistent with the X-ray observations but not the radio limits; 3) The radio and x-ray light curves are consistent with an off-axis relativistic jet with an energy of $10^{49} - 10^{50}$ erg and opening angle of $\theta_j = 15°$ expanding into a low-density medium of $\sim 10^{-4} - 10^{-2}$ cm^{-3}, which we are viewing at $\theta_{obs} \approx 20 - 40°$. We argue that GW170817 is the first detection of an off-axis afterglow from a sGRB. (Margutti *et al.* 2017, Alexander *et al.* 2017).

The sGRB that followed GW170817 was subluminous: it occurred closer than any other sGRB with a known z, and had an energy $1,000\times$ smaller than the next two closest sGRB with known z. (Abbott *et al.* 2017h). Our radio and x-ray observations allow us to place the afterglow in the context of other sGRB afterglows. In the x-ray, the counterpart to GW170817 is $3,000\times$ less luminous than the median of the sGRB population, and $50\times$ less luminous than the previous lowest. In the radio, the countpart's luminosity is $10000\times$ less then the median of the population, and $500\times$ less than the previous lowest. On the other hand, the physical properties of the jet and environment inferred from

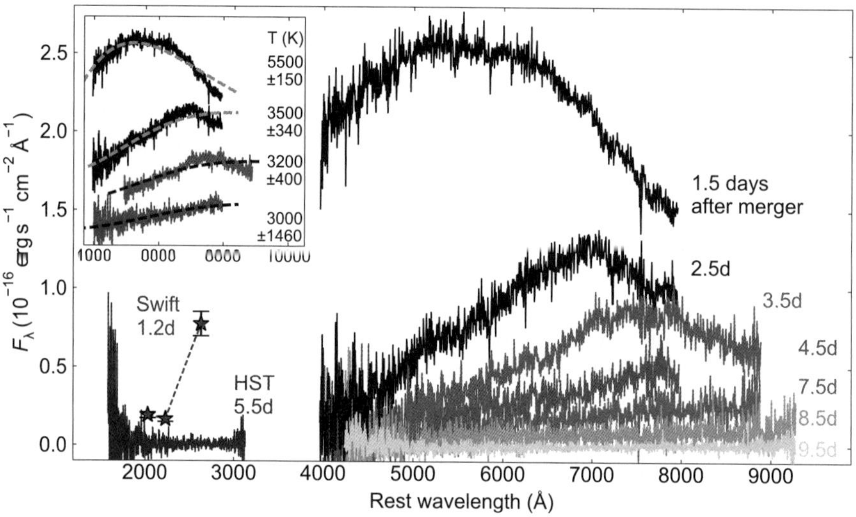

Figure 4. Optical spectroscopy of the counterpart for the first 10 days after merger. (Nicholl *et al.* 2017)

the joint radio and x-ray modeling are strikingly similar to those of the rest of of the sGRB population. The clear inference is that the primary difference between the sGRB associated with GW170817 and the rest of the population is the effect of viewing angle: we are viewing it off-axis. If the counterpart had been at the redshifts usual for sGRB, it would not have been detected with the current generation of radio and x-ray telescopes. (Fong *et al.* 2017).

We are thus able to compare the cosmological population of short GRBs to a local analog. NGC 4993 has higher stellar population age and stellar mass but lower star formation rate then the rest of the sGRB host population. The counterpart to GW170817 is closer to the center of NGC 4993 than 70% of the sGRB population to their hosts. The progenitor of GW170817 may have had a comparatively small kick velocity or perhaps just happened to be near pericenter of its orbit at the time of merger (Fong *et al.* 2017). NGC4993 stands out from the population of sGRB hosts as less asymmetric and more concentrated. sGRB hosts tend to be late type galaxies. Clear major galaxy mergers are unusual amongst sGRB hosts. However, it is entirely plausible that many of them have undergone a recent minor galaxy merger that has increased their asymmetry. (Palmese *et al.* 2017).

5. Observations of the Kilonova

In order to understand the nature of the counterpart we obtained optical spectra using the 4m SOAR & 6.5m Magellan telescopes, and the 2.4m HST STIS instrument (Nicholl *et al.* 2017), and infrared spectra from the 8m Gemini telescope (Chornock *et al.* 2017) of the counterpart to understand the nature of the light emission.

Figure 4 shows the high cadence optical spectra. Spectra at 7.5 days and earlier are from SOAR, those later are from Magellan. The source spectrum at 1.5 days after merger is a featureless near-blackbody of T=5000 K. Over the next 24 hours the source cools by 1500 K, and by day 4.5 the peak of the spectrum is in the infrared, $\lambda > 800$nm. The spectra remain featureless throughout the observation period. The spectra are too red to be consistent the synchrotron emission generally seen in sGRB afterglows and too bright to be consistent with the modeling of the afterglow in Margutti *et al.* 2017. The

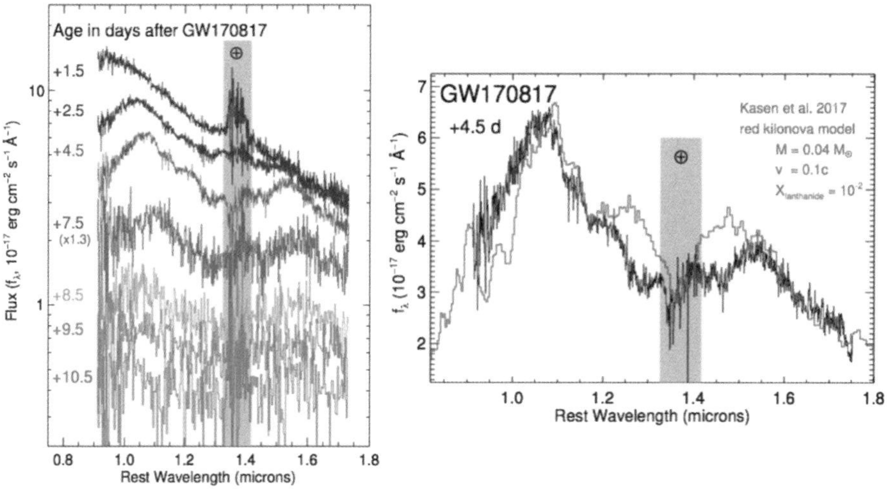

Figure 5. On the left: NIR specta of the counterpart for the first 10 days after merger. On the right, the day 4.5 Gemini NIR spectra with the best fit kilonova model overplotted. (Chornock *et al.* 2017)

spectra are too red to be consistent with supernova spectra, and even supernova with expansion velocities of $\approx 0.1c$ have highly featured spectra. The spectra of the counterpart are notably featureless. The spectra can be modeled by kilonova model spectra (Kasen *et al.*(2013), Kasen *et al.* 2017) if the Lanthanide fraction is very low ($\lesssim 10^{-4}$), to make the spectra sufficiently blue, and the expansion velocities are $\approx$ 0.3c, to remove structure from the spectrum. (Nicholl *et al.* 2017).

Figure 5 shows the high cadence NIR spectra. The source spectrum at 1.5 days after merger is smooth and near-linear. Over the next 24 hours a peak appears at 1μm and by 4.5 days after merger another peak at 1.5μm is prominent. source cools by 1500 K, and by day 4.5 the peak of the spectrum is in the infrared, $\lambda >$ 800nm. This evolution from a smooth blue continuum to a redder continuum with two dominant peaks suggests a transition from one source opacity to another. The day 4.5 spectrum is particularly informative, as seen on right in Figure 5. When we fit the day 4.5 spectrum with models from Kasen *et al.* 2017, we find the best fit model has a high Lanthanide fraction, $\approx 10^{-2}$ and an expansion velocity of 0.1c. All of the spectral models of kilonova, from the early (Kasen *et al.*(2013), Hotokezaka *et al.*(2013)) to the latest (Kasen *et al.* 2017, Tanaka *et al.* 2017) models have the basic features of peaks near 1.1 and 1.55 μm, and agree that these indicate the presence of r-process nucleosynthesis elements as the dominant opacity source, as opposed to a spectral peak in the optical for material dominated by iron-peak opacities. The agreement between model and data is astonishingly good, given the uncertainties in the underlying atomic data and the simplifications inherent to a single-component model. (Chornock *et al.* 2017).

The $i - z$ color and M_i from the first DECam data are consistent with a T=8300 K blackbody with a luminosity of $\lambda L_\lambda \approx 5 \times 10^{41}$ erg/sec. At 0.6 days since merger, this implies an expansion velocity of 0.3c. This is consistent with the optical spectra obtained in the following few days. The infrared spectra require a different component, expanding closer to 0.1c. We fit spectral models of kilonova as a function of ejecta mass, ejecta velocity, and opacity/Lanthanide fraction to light curves of u, g, r, i, z, Y DECam and H, K_s Gemini-S FLAMNGOS-2 photometry over the two weeks subsequent to the merger event using the MOSFit light curve modeling program. Single component models did not describe the data, as expected. The best fitting two component and best fitting three component models were indistinguishable from an information criteria score; Occam's razor suggests we select the two component model. The best fitting model had

- *blue component:* $M_{ej} = 0.014^{+.002}_{-.001}$ $M_{\odot}$, $v/c = 0.266^{+.007}_{-.002}$, & $\kappa \equiv 0.5$ cm^2/gm
- *red component:* $M_{ej} = 0.036^{+.001}_{-.002}$ $M_{\odot}$, $v/c = 0.123^{+.012}_{-.014}$, & $\kappa = 3.35^{+.36}_{-.34}$ cm^2/gm

Matching the derived Galactic r-process production rate requires kilonova ejecta masses in the range of 0.01 - 0.1 $M_{\odot}$ given estimates of the binary NS merger rate from the Galactic neutron star population. Our best fit two-component model has a total ejecta mass of 0.055 $M_{\odot}$, with 0.035 $M_{\odot}$ of material enriched in the heavy r-process, squarely in the middle of the range. These data are clear evidence that binary NS mergers are the site of r-process nucleosynthesis. (Cowperthwaite *et al.* 2017).

The data presented in this set of papers represent by far the best observations of an r-process powered kilonova, and it is remarkable how well the observations match theoretical models.

6. Measuring the Hubble Constant with GW

The GW amplitude from a source is predominately set by $\frac{M}{d}\cos\theta_{obs}$. The chirp mass M can be constrained by the GW frequency and its derivatives, leaving as the main unknowns the distance and inclination angle, θ_{obs}. This standard siren proposal (Schutz(1986) & Holz & Hughes(2005)) is most powerful when the z of a counterpart to the GW event can be measured.

The redshift of NGC 4993 is known; our search of the LIGO/Virgo spatial localization area did not depend on the GW distance or host z; therefore the redshift of the source can be used as an independent variable in the joint cosmological analysis. The z of NGC 4993 is small enough that the Hubble expansion does not dominate. NGC 4993 is part of a small group of galaxies; the average z of the group is a better estimator of the Hubble expansion. The group itself is streaming towards the Great Attractor and a correction for this peculiar velocity is made based on peculiar velocity maps over large parts of the sky. The resulting Hubble expansion is $z = 0.0101 \pm 0.0005$. The Hubble Constant is, for small z, $H_0 = cz/d$. The GW signal from GW170817 was the highest yet, at S/N=32 and the distance constraints are relatively strong. Further, the GW polarization information from having a third GW detector, Virgo, constrains θ_{obs} to $144° - 180°$. The distance measured is $d = 43.8^{+2.9}_{-6.9}$ Mpc. A Bayesian analysis inferred a posterior distribution on H_0 and θ_{obs}, marginalized over uncertainties in z, and found $H_0 = 70^{+12}_{-8}$ km/s/Mpc. Figure 6 shows the result of

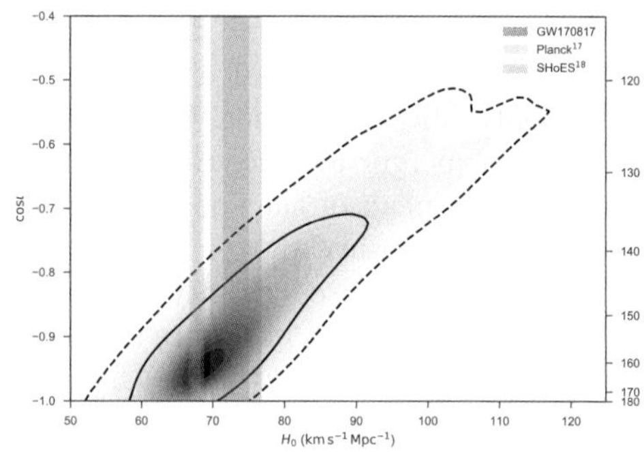

Figure 6. The posterior density for H_0 and $\cos\theta_{obs}$ is shown in blue contours. The brown band is the astrophysical distance ladder constraint on H_0, while the green band is the cosmological inverse distance ladder constraint. (Abbott *et al.* 2017b) Since inception our collaboration has included the Holz LIGO Group at Chicago: Dan Holz, Hsin-Yu Chen, Ben Farr, Zoheyr Doctor.

this analysis, illustrating the GW degeneracy between distance and θ_{obs}. This is the first step to the era of percent-level gravitational-wave cosmology. (Abbott *et al.* 2017b)

If one could place constraints on θ_{obs} from EM observations, the uncertainty on the GW distance can be lessened. We made an investigation of this using by modeling the off-axis jet using the x-ray and radio emission up to 40 days after the merger. Because of the shape of the H_0-$\cos\theta_{obs}$ degeneracy, the result of our analysis, $\theta_{obs} \approx 25° - 50°$, doesn't significantly lessen the uncertainties on the H_0 measurement, but it does shift the maximum a posterior value of H_0 to higher values. (Guidorzi *et al.* 2017).

7. Summary

The DECam on the Blanco 4m telescope was used to both perform an independent discovery of the counterpart to GW170817 and to eliminate other possible candidates in the spatial localization area. Our collaboration studied the host galaxy, the radio and x-ray afterglow, and the lightcurve and optical and NIR spectroscopy of the counterpart. Modeling of the DECam light curve provides a concrete two-component model for the kilonova with 0.01 $M_{\odot}$ ejected with low opacity and high velocity, and 0.04 $M_{\odot}$ ejected with high opacity and low velocity. With the identity of the counterpart secure, the first measurement of H_0 using gravitational wave standard sirens was performed.

References

Abbott, B. P., Abbott, R., Abbott, T. D., *et al. Nature*, 551:85–88, 2017
Abbott, B. P., Abbott, R., Abbott, T. D., *et al. Physical Review Letters*, 119(16):161101, 2017
Abbott, B. P., Abbott, R., Abbott, T. D., *et al. Astrophysical Journal Letters*, 848:L12, 2017
Abbott, B. P., Abbott, R., Abbott, T. D., *et al. Astrophysical Journal Letters*, 848:L13, 2017
Alexander, K. D., Berger, E., Fong, *et al. Astrophysical Journal Letters*, 848:L21, 2017
Allam, S., Annis, J., Berger, E., *et al. GCN* 21530, 2017
Annis, J., Soares-Santos, M., Berger, E., *et al. Astrophysical Journal Letters*, 823:L34, 2017
Barnes, J., & Kasen, D. *Astrophysical Journal*, 775:18, 2013.
Berger, E., *Annual Reviews of Astronomy and Astrophysics*, 52, 43, 2014.
Blanchard, P. K., Berger, E., Fong, *et al. Astrophysical Journal Letters*, 848:L22, 2017
Chornock, R., Berger, E., Kasen, D., *et al. Astrophysical Journal Letters*, 848:L19, 2017
Cowperthwaite, P. S., *et al. Astrophysical Journal Letters*, 826:L29, 2016
Flaugher, B., Diehl, H. T., *et al.*, and DES Collaboration. *Astronomcial Journal*, 150:150, 2015.
Fong, W., Berger, E., Blanchard, P. K., *et al. Astrophysical Journal Letters*, 848:L23, 2017
Guidorzi, C., Margutti, R., Brout, D, *et al. ArXiv e-prints*, October 2017.
Holz, D. E., & Hughes, S. A. *Astrophysical Journal*, 629:15, 2005.
Hotokezaka, K., Kiuchi, K., Kyutoku, K., *et al. Physical Review D*, 87:024001, 2013.
Kasen, D., Metzger, B., Barnes, J., *et al. Nature*, 551:80–84, November 2017.
Kasen, D., Badnell, N. R., & Barnes, J. *Astrophysical Journal*, 774:25, 2013.
Margutti, R., Berger, E., Fong, W., *et al. Astrophysical Journal Letters*, 848:L20, 2017
Metzger, B. D., & Berger, E. *Astrophysical Journal*, 746:48, 2016.
Nicholl, M., Berger, E., Kasen, D., *et al. Astrophysical Journal Letters*, 848:L18, 2017
Palmese, A., Hartley, W., Tarsitano, *et al. Astrophysical Journal Letters*, 849:L34, 2017
Schutz, B. F. *Nature*, 323:310, 1986.
Soares-Santos, M., Kessler, R., Berger, E., *et al. Astrophysical Journal Letters*, 823:L33, 2016
Soares-Santos, M., Holz, D. E., Annis, J., *et al. Astrophysical Journal Letters*, 848:L16, 2017
Tanaka, M., Utsumi, Y., Mazzali, P. A., *et al. ArXiv e-prints*, October 2017.

Gravitational Wave Astrophysics: Early Results from Gravitational Wave Searches and Electromagnetic Counterparts
Proceedings IAU Symposium No. 338, 2017
G. González & R. Hynes, eds.

doi:10.1017/S1743921318003630

Observations of the first electromagnetic counterpart to a gravitational wave source by the TOROS collaboration

Lucas M. Macri[1], Mario C. Díaz[2], Diego Garcia Lambas[3], and the TOROS collaboration

[1]George P. and Cynthia W. Mitchell Institute for Fundamental Physics & Astronomy, Department of Physics & Astronomy, Texas A&M University, College Station, TX, USA
email: lmacri@tamu.edu

[2]Center for Gravitational Wave Astronomy and Department of Physics & Astronomy, University of Texas - Río Grande Valley, Brownsville, TX, USA

[3]IATE-OAC, Universidad Nacional de Córdoba-CONICET, Córdoba, Argentina

Abstract. We present the results of prompt optical follow-up of the electromagnetic counterpart of GW170817 by the Transient Optical Robotic Observatory of the South Collaboration (TOROS). We detected highly significant dimming in the light curves of the counterpart over the course of only 80 minutes of observations obtained ∼35 hr after the trigger with the T80-South telescope. A second epoch of observations, obtained ∼59 hr after the event with the EABA 1.5m telescope, confirms the fast fading nature of the transient. The observed colors of the counterpart suggest that this event was a "blue kilonova" relatively free of lanthanides.

Keywords. Gravitational waves, stars: neutron

1. Introduction

The Transient Optical Robotic Observatory of the South Collaboration (TOROS; Díaz *et al.* 2014) was organized in 2013 to participate in electromagnetic observations of gravitational-wave events recorded by LIGO and Virgo. While seeking to deploy a wide-field optical telescope on Cordón Macón (NW Argentina), we have used other resources for follow-up activities (Colazo *et al.* 2015, Díaz *et al.* 2016).

2. Observations

On 2017 August 18 (∼35 hr after the trigger of GW170817; Abbott *et al.* 2017) we observed the EM counterpart (Coulter *et al.* 2017) using the 0.8-m T80-South telescope (T80-S) at CTIO. We used its 85 Mpix camera (FoV of 1.4° on a side) to obtain 16, 15, and 15 one-minute exposures through SDSS *gri* filters, respectively, over the course of 80 minutes. The left panel of Fig. 1 shows a color composite of a small sub-section of the T80S FoV centered on the transient.

On 2017 August 19 (∼59 hr after the GW trigger) we imaged the source with the 1.54-m telescope at the Estación Astrofísica de Bosque Alegre (EABA) and a 17 Mpix camera with a FoV of 17′ on a side. We obtained 88 one-minute unfiltered exposures.

3. Photometry

We subtracted NGC 4993 from all of the images by fitting a Sérsic profile with IMFIT (Erwin 2015). The right panel of Fig. 1 shows the result. We carried out time-series

Figure 1. Left: pseudo-color image of a small subsection (2.4′ on a side) of the FoV of T80-S, centered on the transient (marked with crosshairs). Intensity scaling is logarithmic in order to better display the light distribution of the host galaxy. Right: residual image after host galaxy subtraction and core masking (hatched circle).

PSF photometry of the T80-S images using DAOPHOT, ALLSTAR, and ALLFRAME (Stetson 1987, Stetson 1994) following the procedures described in Macri *et al.* (2015).

We identified 100 bright & isolated stars in each filter within the T80-S FoV and solved for frame-to-frame offsets to correct for differential extinction and any other variations. We achieved a precision in our time-series photometry of 0.01 mag or better for objects with $g < 16$, $r < 15$, $i < 14$ mag.

We transformed the T80-S measurements into the Pan-STARRS1 photometric system (Tonry *et al.* 2012) and simultaneously corrected for atmospheric extinction using 4600−5400 stars in common between our star lists and the PS1 catalog. The calibrated light curves are plotted in Fig. 2. The photometric measurements are reported in Table 1 of Diaz *et al.* (2017), which contains additional details of our observations and analysis.

Due to the lower S/N of the EABA observations, we combined those frames to obtain a mean PSF magnitude. We transformed it to PS1 r using 200 stars in common.

4. Analysis

The T80-S light curve of the transient exhibits a very significant decline across all bands during the ∼80 min of observations. A weighted linear fit to the data yields $\Delta g = 0.17 \pm 0.03$ mag, $\Delta r = 0.14 \pm 0.02$ mag, $\Delta i = 0.10 \pm 0.03$ mag over that time period. The mean magnitudes at the mid-point of our observations (1.467 days after the GW trigger) and their time derivatives are:

$$\begin{aligned} g &= 18.60 \pm 0.02 \text{ mag}, & dg/dt &= 3.0 \pm 0.6 \text{ mag/day} \\ r &= 17.99 \pm 0.02, & dr/dt &= 2.5 \pm 0.4 \\ i &= 17.80 \pm 0.02, & di/dt &= 1.9 \pm 0.5 \end{aligned}$$

As we do not have a measurement of $r-i$ during the EABA observations, we present two possible values for its calibrated magnitude at that time (2.456 days after the GW trigger): $r = 18.78 \pm 0.03$ mag if its color did not evolve over 24 hr or $r = 19.15 \pm 0.06$ mag

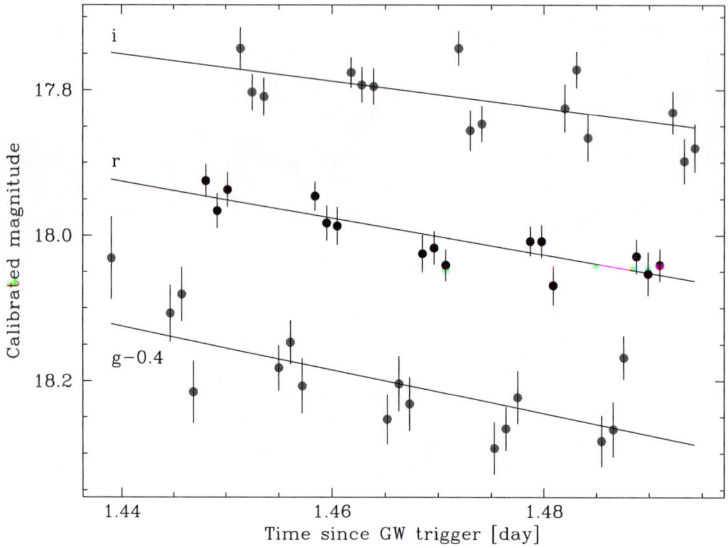

Figure 2. *gri* light curves of the EM counterpart to GW170817, obtained with T80-S on 2017 August 18. The g points have been offset by -0.4 mag for clarity.

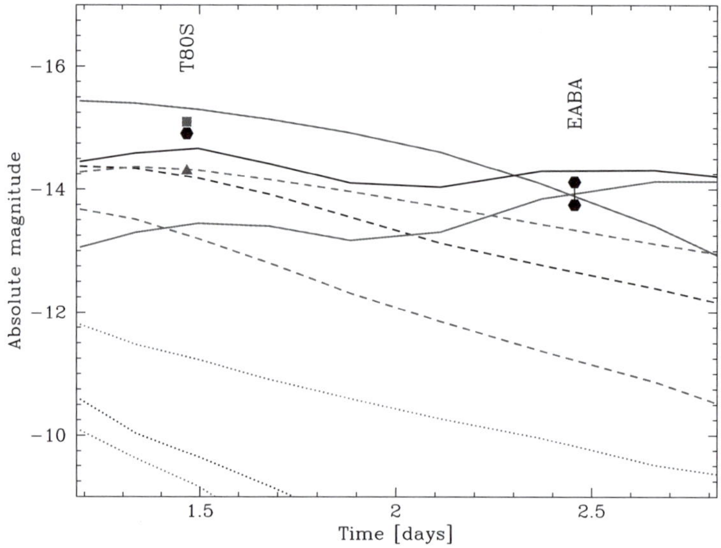

Figure 3. Comparison of our photometry corrected for $D = 38$ Mpc (g: blue; r: black; i: red) with models from Tanaka *et al.* (2018). The dotted lines represent a "red kilonova" model with dynamical ejecta rich in lanthanides. The dashed and solid lines represent "blue kilonova" wind models with decreasing amounts of lanthanides. The measurement uncertainties are smaller than the size of the symbols. The two possible r values at 2.456 days are described in the text.

if its color evolved as extrapolated from the T80-S light curves to $r{-}i \sim 0.84 \pm 0.08$ mag. Despite these limitations, the EABA observations confirm a fast decline of $\Delta r = 0.8 - 1.2$ mag over 24 hr.

Fig. 3 compares our colors and absolute magnitudes (adopting $D = 38 \pm 5$ Mpc, Kourkchi & Tully 2017) with models from Tanaka *et al.* (2018). The measurements are inconsistent with the predictions for a "red kilonova" containing dynamical ejecta

(0.01 $M_{\odot}$, $v = 0.2c$) rich in lanthanides (dotted lines). Our r-band luminosity is in fairly good agreement with the prediction for a "blue kilonova" (Metzger 2017) model with a wind (0.01 $M_{\odot}$, $v = 0.05c$) free of lanthanides ($Y_e = 0.3$ variant, solid lines), but the predicted $g-i$ color at $t = 1.5$ days does not match the observations. A similar model in which the ejecta contain a small amount of lanthanides ($Y_e = 0.25$ variant, dashed lines), matches the observed colors fairly well, but the predicted luminosities are somewhat lower than observed. Regardless, it appears that this event may have been relatively free of lanthanides.

References

Abbott, B. P., *et al.* 2017, *PRL* 116, 061102
Colazo, C., *et al.* 2015, *GRB Coordinates Network, Circular Service,* 18338
Coulter, D. A., *et al.* 2017, *GRB Coordinates Network, Circular Service,* 21529
Díaz, M. C., *et al.* 2014, in *Third "Hot-wiring the Transient Universe" Workshop*, 225
Díaz, M. C., *et al.* 2016, *ApJL* 828, 16
Diaz, M. C. *et al.* 2017, *ApJL* 848, 29
Erwin, P. 2015, *ApJ* 799, 226
Kourkchi, E. & Tully, R. B. 2017, *ApJ* 843, 16
Macri, L. M., *et al.* 2015, *AJ* 149, 117
Metzger, B. D. 2017, *LRR* 20, 3
Stetson, P. B. 1987, *PASP* 99, 191
Stetson, P. B. 1994, *PASP* 106, 250
Tanaka, M., *et al.* 2018, *ApJ* 852, 109
Tonry, J. L., *et al.* 2012, *ApJ* 750, 99

Gravitational Wave Astrophysics: Early Results from Gravitational Wave Searches and Electromagnetic Counterparts
Proceedings IAU Symposium No. 338, 2017
G. González & R. Hynes, eds.
doi:10.1017/S1743921318003605

AGILE Observations of GW Events

Francesco Verrecchia[1,2], Marco Tavani[3,4,5], and Immacolata Donnarumma[6,3], on behalf of the AGILE Team

[1]ASI Space Science Data Center (SSDC), via del Politecnico, I-00133, Roma, Italy
email: francesco.verrecchia@ssdc.asi.it

[2]INAF-OAR, via Frascati 33, I-00078, Monte Porzio Catone (Roma), Italy

[3]INAF-IAPS, via del Fosso del Cavaliere 100, I-00133 Roma, Italy

[4]Dipartimento di Fisica, Università di Roma "Tor Vergata", via della Ricerca Scientifica 1, I-00133 Roma, Italy

[5]Gran Sasso Science Institute, viale Francesco Crispi 7, I-67100 L'Aquila, Italy

[6]ASI, via del Politecnico snc, I-00133, Roma, Italy

Abstract. AGILE is a space mission of the Italian Space Agency dedicated to γ-ray astrophysics, launched in 2007. AGILE performed dedicated real-time searches for possible γ-ray counterparts of gravitational wave (GW) events detected by the LIGO-Virgo scientific Collaboration (LVC) during the O2 observation run. We present a review of AGILE observations of GW events, starting with the first, GW150914, which was a test case for future searches. We focus here on the main characteristics of the observations of the most important GW events detected in 2017, i.e. GW170104 and GW170817. In particular, for the former event we published γ-ray upper limits (ULs) in the 50 MeV – 10 GeV energy band together with a detailed analysis of a candidate precursor event in the Mini-Calorimeter data. As for GW170817, we published a set of constraining γ-ray ULs obtained for integrations preceding and following the event time. These results allow us to establish important constraints on the γ-ray emission from a possible magnetar-like remnant in the first ~ 1000 s following T_0. AGILE is a major player in the search of electromagnetic counterparts of GW events, and its enhanced detection capabilities in hard X-ray/MeV/GeV ranges will play a crucial role in the future O3 observing run.

Keywords. gravitational waves, gamma rays: general

---

1. Introduction

The first historical discovery of a gravitational wave (hereafter, GW) detected by the LIGO detector from the source GW150914 (Abbott *et al.* 2016), opened the era of GW detection from astronomical sources. Furthermore, the most recent discovery of the first electromagnetic (e.m.) counterpart of the source GW170817/GRB170817A, obtained with the contribution of the Virgo detector, definitely started the era of multi-messenger astronomy with GWs. AGILE is a space mission of the Italian Space Agency dedicated to γ-ray and X-ray astrophysics, in a near equatorial orbit since 2007. It consists of an imaging gamma-ray Silicon Tracker (sensitive in the energy range 30 MeV – 30 GeV), a coded mask hard X–ray imager Super-AGILE (SA; 20 – 60 keV), a Mini-Calorimeter (MCAL; 0.4 – 100 MeV) and an anticoincidence (AC) system (see Tavani *et al.* 2009). The combination of Tracker, MCAL and AC working as a gamma-ray imager constitutes the AGILE-GRID. The instrument is capable of detecting γ-ray transients and GRB-like phenomena for timescales ranging from sub-milliseconds to tens-hundreds of seconds.

Table 1. AGILE observations of LVC GW events.

GW ID	GCN #	Perc.[1]	Δ t [T - T_0]	Comments
150914	—	0	+330	LR missed by 300 s; earliest ever *gamma*-ray
151226	—	30	0	data on GW event; Tavani *et al.* 2016
170104	20375, 20395	36	0	partially exposed by GRID, fully exp. by MCAL; Verrecchia *et al.* 2017a
170608	21224, 21228	40	0	partially exposed by GRID, exposed by MCAL
170814	21477, 21482	0	+500	exposed after 500 s (1st with Virgo data); —
170817	21525, 21526 21562, 21785	0	+930	exposed after 930 s; prompt OT not exposed; limits to magnetar-like emission; Verrecchia *et al.*2017c

Notes:
[1] Percentage of the prompt AGILE-GRID exposure of the 90% c.l. LVC LR.

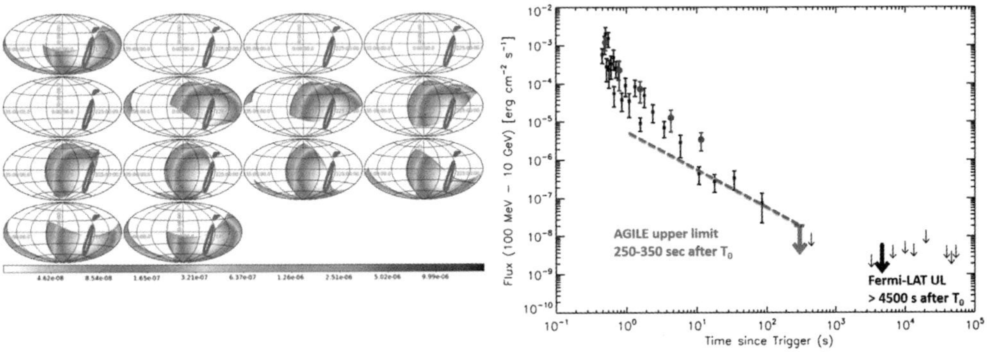

Figure 1. The sequence of UL maps for passes over the GW150914 LR preceeding and following T_0, with the best exposure of the LVC LR occurred at +330 s (left panel). The AGILE-GRID 2σ UL in the 50 MeV – 10 GeV compared with the GRB 090510 AGILE-GRID and Fermi/LAT light curves rescaled at the GW redshift 0.1 (right panel).

Moreover, with the standard maximum likelihood analysis, AGILE routinely detects gamma-ray transients varying from time scale of $\sim$ hours to weeks.

The AGILE Team signed a Memorandum of Understanding with the LIGO-Virgo Collaboration (LVC) on November 2016. We review the AGILE observations and the dedicated activities for the GW γ-ray counterpart search.

2. Overview

AGILE observations of the main LVC GW events detected since 2015 are summarized in Table 1. We report in the table the time difference $\Delta\, t$ between the AGILE observations and the GW trigger time (T_0). At time $T = T_0 + \Delta\, t$ the AGILE-GRID Field of View overlapped the LVC 90% c.l. localization region (LR), covering a fraction reported in the table. We show that in three events we had a partial coverage of the LR at event T_0.

Taking into account the AGILE spinning operative mode, we considered the sequence of exposure maps for 100 s integrations preceeding and following the first GW event GW150914 (see Fig. 1). All integrations are independent and consecutive, and if combined to an integration time of about 7 minutes, we obtained a coverage of 80% of the sky.

The first GW event, whose analysis was executed after the event public announcement (about five months after the discovery), was then used to plan the on-ground dedicated automatic pipelines. We report below the main results obtained for this event:

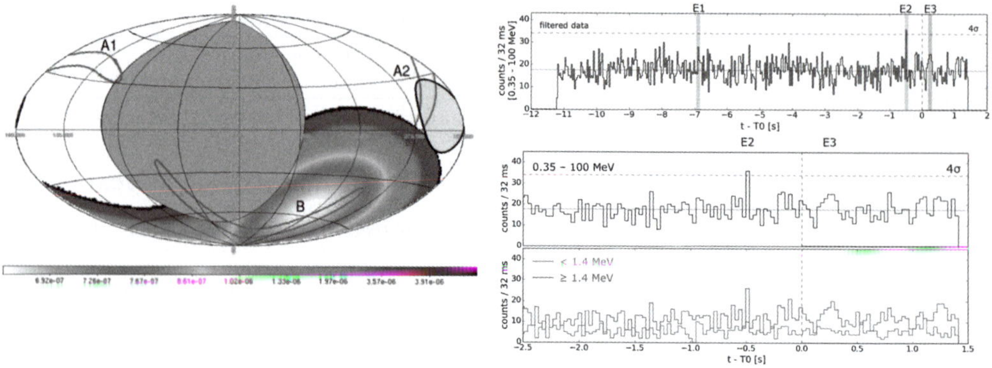

Figure 2. The GRID $3\,\sigma$ UL map (in 50 MeV – 10 GeV) on 4 s integration around GW170104 T_0 showing a partial spatial coverage of the event LR (left panel). MCAL 12.6 s photon-by-photon data, including T_0 (right top panel), showing three flagged times of which E2 is a candidate precursor event. A smaller time interval around E2 in two energy bands is also shown (right bottom panel).

GW150914: the event occurred at $T_0 =$ 09:50:45 UTC, 2015 September, 14. AGILE analysis started after event publication on 2016 February 11, with the execution of an archival search. We missed the coverage of the prompt event just for few tens of seconds. We obtained a good coverage of the 90% c.l. LR with the GRID at +330 s (time with respect to T_0) with an integration of 100 s (Tavani *et al.* 2016). We determined a $2\,\sigma$ γ-ray upper limit (UL) of $1.9 \times 10^{-8}\,\mathrm{erg\,cm^{-2}\,s^{-1}}$ in the 50 MeV – 10 GeV energy band. This UL is compared in Fig. 1 with the GRID and Fermi/LAT light curves of the short GRB 090510 rescaled at z = 0.1. This UL was the earliest available measurement in the γ-ray band for a GW event. An extrapolation of the GRID sensitivity at T_0 green dotted curve in Fig. 1) shows that AGILE might be able to detect prompt γ-ray emission similar to that of GRB 090510. In any case the AGILE early data near +300 s are very significant.

2.1. *Learned lesson: AGILE preparation for O2 in 2016*

In preparation for the LIGO O2 run, a reconfiguration of the AGILE instrument was carried out in August 2016. The activity included an improved MCAL burst detection with a lower threshold to allow triggering from the "sub-threshold" events (assumed to be similar to the GRB 090510 precursor event). Moreover, an AGILE-GW LVC notice/Circular reaction pipeline was developed to improve the performance of MCAL and GRID data processing at the publication of an LVC notice or circular, and to alert members of the AGILE-GW flare advocate group active round-the-clock.

3. Recent results: AGILE observations during O2 run

We report below the results of the observations of the main GW events detected during O2.

GW170104: the event occurred at $T_0 =$ 10:11:58.59 UTC, 2017 January 4. AGILE analysis started soon after the LVC circular publication (Abbott *et al.* 2017a) and two GCN on MCAL and GRID data analysis were published (Tavani *et al.* 2017a, Tavani *et al.* 2017b) reporting that at T_0 the 90% LR was partially covered by AGILE (Fig. 2). A dedicated paper was later published, reporting the main analyses of AGILE detectors data (Verrecchia *et al.* 2017a). In particular, MCAL photon-by-photon data were

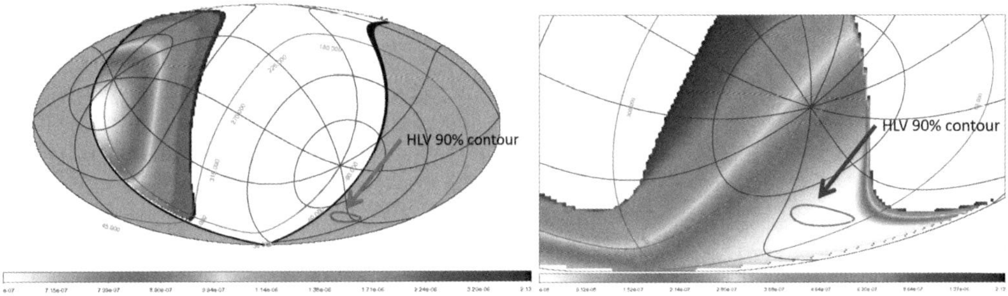

Figure 3. The GRID 3 σ UL map (in 50 MeV – 10 GeV) on 4 s integration around the GW170814 T_0 showing Earth occultation of the event LR (left panel). The 100 s best exposure of the LR at +550 s after T_0, with a complete coverage (right panel).

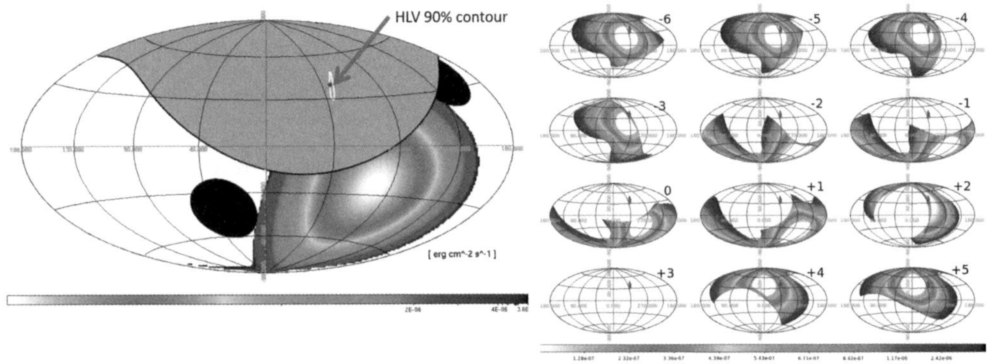

Figure 4. The GRID 2 σ UL map (in 30 MeV – 10 GeV) on 4 s integration around GW170817 T_0, showing the LR occulted by the Earth (left panel). The 150 s exposure UL map scan preceding and following T_0 (right panel) evaluated at the optical transient position.

acquired over a time window of 12.6 s including T_0 due to an onboard trigger. However, after a refined filtering of data, the original trigger was found to be spurious, while a candidate event was detected on the 32 ms time binning in full energy band at -0.5 s. This possible "precursor" event has a pre-trial significance of 4.4 σ. After considering its "false-alarm" rate ($FAR_{E2} = 1 \times 10^{-4}$ Hz), the post-trial significance resulted to be between 2.4 and 2.7 σ.

Additional learned lesson: the analysis of MCAL data of the first event, led to the development of a new automatic detection MCAL pipeline, using various binning time scales from 8 to 128 ms. This procedure has integrated the functionality of the past GRB and TGF pipelines, together with the new one dedicated to "sub-threshold" events.

GW170814: the event occurred at T_0 = 10:30:43 UTC, 2017 August 14, the first event with a crucial contribution to the localization by the Virgo detector. AGILE analysis started immediately after the LVC circular notification, and two GCN on MCAL and GRID data analysis were published a few hours later (Abbott *et al.* 2017b, Longo *et al.* 2017a, Longo *et al.* 2017b).

The event LR, much smaller than previous ones because of Virgo detector data, was occulted by the Earth at the T_0 (see Fig. 3). The earliest complete coverage of the event

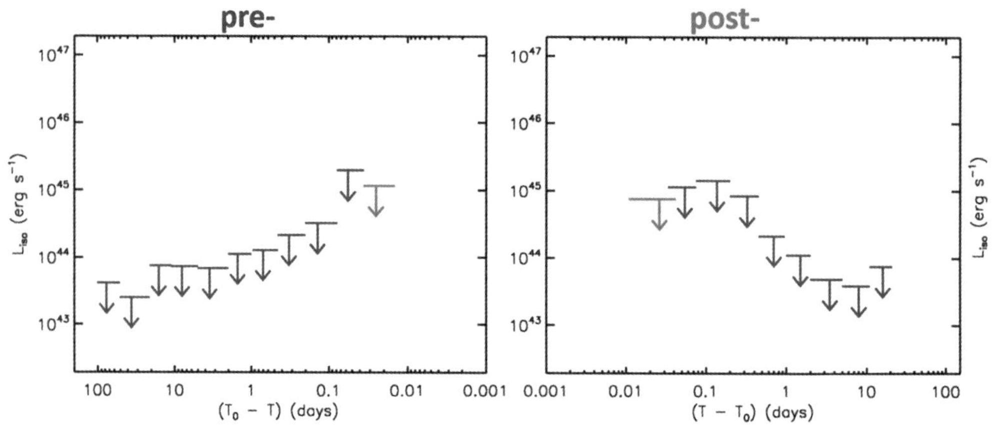

Figure 5. The GRID $2\,\sigma$ luminosity limits (in 30 MeV – 10 GeV) with integrations ranging from 150 s to 50 days preceding and following (right panel) GW170817 T_0. The UL values on 150 s exposures nearest in time to T_0, preceding and following it, are shown in red.

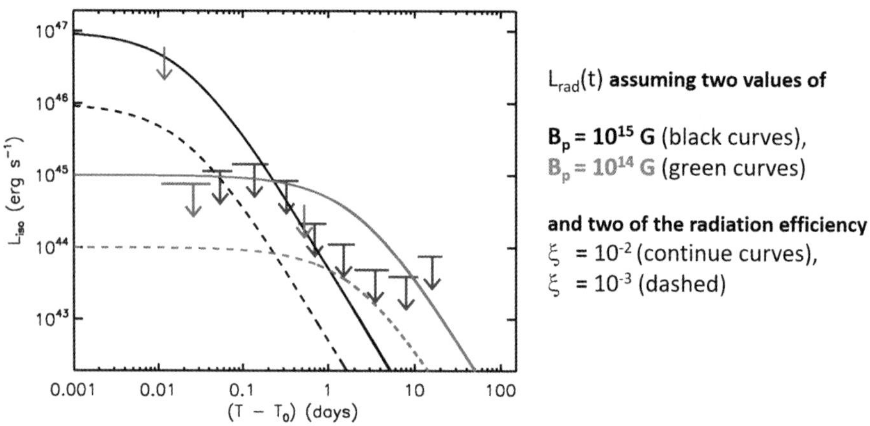

Figure 6. The isotropic luminosity limits (in erg s^{-1}) derived from flux ULs following GW170817 T_0 for the OT distance of 40 Mpc. The AGILE-GRID ULs in 30 MeV – 10 GeV (red and blue arrows), and the SA and MCAL ones (magenta and gray) in the 20 – 60 keV and 400 keV – 100 MeV energy bands, respectively. We also show the high-energy luminosity curve relative to the magnetar-like remnant model described in the text.

LR was been obtained with a 100 s exposure centered at +550 s, producing a $3\,\sigma$ γ-ray UL of 3.4×10^{-8} erg cm^{-2} s^{-1} in the 50 MeV – 10 GeV energy band.

GW170817: the event occurred at $T_0 =$ 12:41:06 UTC, 2017 August 17, together with the detection of GRB 170817A by Fermi/GBM (Abbott *et al.* 2017c, Abbott *et al.* 2017d).
AGILE analysis started soon after the LVC circular submission, and two GCN on MCAL and GRID data analysis were then published. Later on, two more GCNs on GRID data analysis exposing the position of the optical transient (OT) were published (Pilia *et al.* 2017, Piano *et al.* 2017, Bulgarelli *et al.* 2017, Verrecchia *et al.* 2017b). The event LR was occulted by the Earth at T_0 (see Fig. 4), and the earliest exposure was obtained +930 s after T_0. Furthermore, several time windows before and after T_0 were explored to search for possible precursor and/or delayed emissions. GRID data analysis was executed on three time scales with two different methods (Verrecchia *et al.*

2017c), with increasing integration times from 150 s to 50 days. We did not detect a significant emission (although two low significant detections were reported in Verrecchia *et al.* 2017c). This exposure scan allowed to obtain an interesting set of $2\,\sigma$ flux ULs in the 30 MeV – 10 GeV (see Fig. 5). The SA hard X–ray imager had useful data starting at +0.53 d after T_0. The γ-ray flux ULs converted to luminosity limits, assuming the OT distance of 40 Mpc, were applied to a magnetar model (Fig.6), (i.e., a rapidly rotating NS remnant with an high poloidal magnetic field B_p). We assume the γ-ray emission to be proportional to the magnetar spin-down luminosity with conversion to radiation efficiency ξ. AGILE γ-ray ULs are particularly constraining in the first thousand of s after T_0 for $B_p = 10^{15}$G and $\xi = 10^{-2}$.

4. Summary

AGILE interestingly observed all the GW events detected by the LIGO detectors as well as the recent ones detected also by Virgo in the O2 run. Considering the two confirmed 2015 events and all the confirmed O2 ones, we had partial coverage at T_0 for 50% of the events. The results for GW170817/GRB 170817A and those obtained on MCAL data for GW170104, have shown the importance of enhanced on ground activities, such as an automatic MCAL detection pipeline to issue alerts to LVC. Also important will be the continuous onboard TM acquisition of the hard X–ray SA data, which could *detect and localize* X–ray counterparts similar to GRB 170817A.

AGILE is expected to play a crucial role in O3 in the search for e.m. counterparts of GW events. Its unique detection and imaging capabilities in the hard X-ray/MeV/GeV ranges are special assets for future searches. A dedicated effort is ongoing to improve on-board and on-ground detection, processing and automatic alerting procedures. AGILE is well equipped for unveiling the exciting phenomena of this New Astronomy.

References

Abbott, B. P., Abbott, R., Abbott, T. D., *et al.* 2017a, *PhRvL*, 118, 221101

Abbott, B. P., Abbott, R., Abbott, T. D., *et al.* 2017b, *PhRvL*, 119, 141101

Abbott, B. P., Abbott, R., Abbott, T. D., *et al.* 2017c, *PhRvL*, 119, 161101 (A17b)

Abbott, B. P., Abbott, R., Abbott, T. D., *et al.* 2017d, ApJL, 848, L12 (MMA17)

Bulgarelli, A., Tavani, M., Verrecchia, F., *et al.* 2017, GRB Coordinates Network, 21564

Longo, F., Piano, G., Cardillo, M., *et al.* 2017a, GRB Coordinates Network, 21477, 1

Longo, F., Verrecchia, F., Piano, G., *et al.* 2017b, GRB Coordinates Network, 21482, 1

Piano, G., Verrecchia, F., Pilia, M., Cardillo, C., Tavani, M., *et al.* 2017, GRB Coordinates Network, 21526

Pilia, M., Cardillo, C., Piano, G., *et al.* 2017, GRB Coordinates Network, 21525

Tavani, M., Barbiellini, G., Argan, A., *et al.* 2009, A&A, 502, 995

Tavani, M., Pittori, C., Verrecchia, F., *et al.* 2016, ApJL, 825, L4

Tavani, M., Ursi, A., Fuschino, F., *et al.* 2017a, GRB Coordinates Network, 20375, 1

Tavani, M., Verrecchia, F., Minervini, G., *et al.* 2017b, GRB Coordinates Network, 20395, 1

Verrecchia, F., Tavani, M., Ursi, A., *et al.* 2017a, ApJl, 847, L20

Verrecchia, F., Cardillo, M., Bulgarelli, A., *et al.* 2017b, GRB Coordinates Network, 21785

Verrecchia, F., Tavani, M., Donnarumma, I., *et al.* 2017c, ApJl, 850, L27

Gravitational Wave Astrophysics: Early Results from Gravitational Wave Searches and Electromagnetic Counterparts
Proceedings IAU Symposium No. 338, 2017
G. González & R. Hynes, eds.

doi:10.1017/S1743921318003824

High power in Advanced LIGO

Terra Hardwick
for the LIGO Scientific Collaboration
Louisiana State University,
Baton Rouge, LA 70803
email: terra.hardwick@ligo.org

Abstract. The LIGO detectors have just completed a successful and exciting observation run. Both facilities are now undergoing upgrades and commissioning, including doubling the circulating power in the interferometer which will increase LIGO's sensitivity above 100 Hz. This paper motivates the power increase and discusses the problems in general that arise with higher power and the progress to date with addressing them. Topics include input power noise coupling, parametric instability, and thermal effects.

Keywords. gravitational waves, instrumentation: detectors, instrumentation: adaptive optics

1. Introduction

On September 14, 2015 LIGO made the first ever detection of gravitational waves. This and the next three detections were of binary black hole systems with component masses between 7.5 - 36 solar masses. On August 17, 2017 LIGO detected gravitational waves from a binary neutron star system for which electromagnetic counterparts were also detected. The final neutron star coalescence occurred outside of LIGO's sensitivity band and several of the black hole mergers were pushing that boundary. Better sensitivity at higher frequencies would result in increased range, higher signal to noise ratio (SNR) for all detections and a greater chance of seeing the full merger and ringdown of more systems. These imply better sky localization, better parameter estimation, better characterization of post-merger activity, and a better understanding of the neutron star equation of state.

During LIGO's first two observing runs (O1, O2), the detector operated with 25 W input power; we plan to double this power for the third observing run scheduled to begin Fall 2018. Increasing the power in the interferometer reduces photon shot noise - currently a limiting noise source - which will increase LIGO's sensitivity in the region above 100 Hz.

High power requires a new laser source scheme designed to integrate into the preexisting LIGO laser layout, provide at least 50 W into the interferometer, and meet low noise requirements. During Fall 2017, this scheme was tested and successfully implemented.

Higher power in the interferometer brings parametric instabilities (PI) and thermal aberrations. While schemes for mitigating these problems were already in place during 02, they must be adapted to the increased power levels.

2. Fundamental noise in LIGO detectors

The sensitivity of LIGO is limited by anything that causes differential length motion; the sum of all such noises is the strain sensitivity curve for the detectors. Fundamental noise sources are those for which there is a theoretical limit given LIGO's parameters. They include noises such as quantum fluctuations, thermal motion, and residual gas noise. Quantum noise consists of variation in photon arrival time - shot noise - at higher

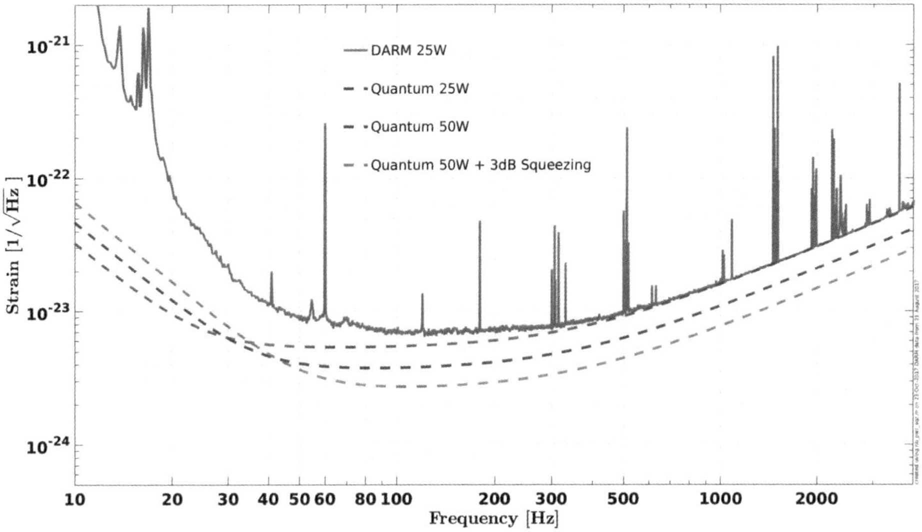

Figure 1. Projection of quantum noise with a) 25 W input power, as during 02, b) 50 W input power, and c) 50 W input power + 3dB squeezing, as expected during 03. Shown alongside differential arm (DARM) strain at LIGO Livingston during O2.

frequencies and variation in photon amplitude - radiation pressure - at low frequencies. Saulson (2017)

Photon shot noise Fluctuations in photon arrival time result in fluctuations of phase which read out as fluctuations of output power on the photodiode. One can also think of this as sensing noise, or how precisely we are able to measure. This photon shot noise is Poisson in nature, causing power fluctuations of

$$\sigma_{power} = \sqrt{n_{photons}} = \sqrt{\frac{\lambda\tau}{4\pi\hbar c}P} \tag{2.1}$$

where τ is the interval over which the number of photons is averaged. In a Michelson interferometer, this fluctuation in power - calibrated in effective strain - has amplitude spectral density

$$h_{shot,FP}(f) = \frac{1}{L}\sqrt{\frac{\hbar c\lambda}{2\pi P}}. \tag{2.2}$$

Note that this is white noise, independent of frequency, but when a full Fabry-Pérot interferometer response is considered, power is frequency dependent. The presence of this photon shot noise puts a fundamental sensitivity limit on strain readout above 100 Hz.

Radiation pressure As photons hit the test mass, they exert a force on it, imparting momentum and causing a recoil reaction. Variations in photon arrival time (as discussed above) cause variations in this force:

$$\sigma_{force} = \frac{1}{c}\sigma_{power} \tag{2.3}$$

This can also be thought about as variation in photon amplitude. The fluctuations in force variable move the test mass, injecting noise into the final length readout. This radiation pressure noise has amplitude spectral density

$$h_{rp}(f) = \frac{1}{mLf^2}\sqrt{\frac{\hbar P_{in}}{2\pi^3 c\lambda}}. \tag{2.4}$$

Note that this is frequency dependent, where the slope is set by the pendulum response of the test mass suspension ($1/f^2$). Radiation pressure puts a fundamental sensitivity limit on strain readout below 100 Hz.

3. Reducing quantum noise

Radiation pressure noise and shot noise can be considered together as one frequency-dependent quantum noise source:

$$h_q(f) = \sqrt{h^2_{shot}(f) + h^2_{rp}(f)}. \tag{3.1}$$

Note that at low frequencies, the $1/f^2$ dependent radiation pressure will dominate, but will be less significant than white shot noise as higher frequencies.

There are two ways to influence quantum noise: squeezing (in amplitude or phase) and changing input power. Squeezing in phase and increasing power will lower LIGO's noise above 100 Hz, giving the desired frequency sensitivity previously discussed. Both techniques add noise at lower frequencies, but currently the LIGO detectors have noise sources (i.e. angular controls and seismic) that are higher than radiation pressure noise so neither scheme will reduce sensitivity at low frequency.

Squeezing Radiation pressure noise and shot noise are most fundamentally explained by quantum vacuum fluctuation. These vacuum fluctuations have an amplitude and phase relationship:

$$\sigma_{amp}\sigma_{phase} \geqslant \frac{\hbar}{2} \tag{3.2}$$

such that there is a limit on the combined decrease of amplitude and phase noise. The quantum noise floor is where $\sigma_{amp}\sigma_{phase} = \frac{\hbar}{2}$; the combined fluctuations of amplitude and phase are as low as possible. Squeezing works on the principle that either can be reduced at the cost of raising the other: phase noise (shot noise) can be lowered if amplitude noise (radiation pressure noise) is raised and vice versa.

To accomplish this, LIGO plans to inject squeezed light into the interferometer. Since LIGO is currently limited by only quantum shot noise at high frequency, phase squeezing will be implemented for the next observing run. While this will subsequently increase amplitude fluctuations, the increased low frequency quantum noise will still sit below other limiting noise factors. Eventually frequency dependent squeezing will be implemented, injecting phase squeezed light at high frequencies and amplitude squeezed light at low frequencies.

Increased power Strain optical gain (unit Watts) is how differential arm strain converts to power on the output photo diode.

$$\text{strain optical gain} := \frac{dP_{out}}{dh} = \frac{8\pi}{c}\phi_0 g_\phi f P_{in} \tag{3.3}$$

where ϕ_0 is frequency offset from full dark. From this we see that optical gain scales linearly with input power: the higher P_{in}, the greater change in P_{out} for differential arm length change. From eq. 2.1, shot noise SNR scales only with the square root of power. Put another way, the fractional fluctuation in number of photons arriving at a test mass

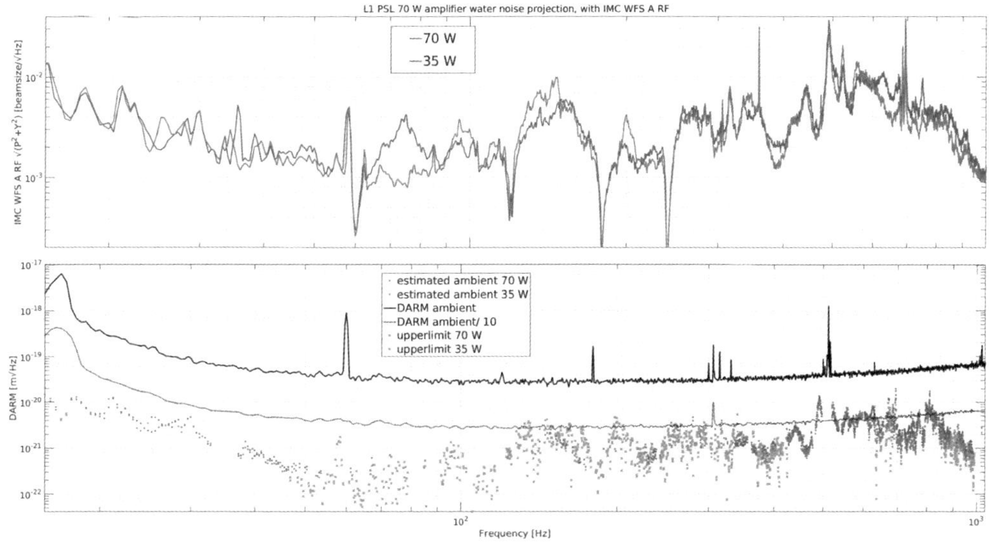

Figure 2. Noise projections of table-motion jitter from previous (orange) and new (blue) laser set ups. First, coupling functions are made by injecting noise until a response is seen in DARM (in this case, shaking the optic table). That function can then be multiplied by ambient noise to estimate how far below DARM the nominal noise sits. If the original injection was not strong enough to couple into DARM, only upper limits can be placed.

in some time τ is

$$\frac{\sigma_{power}}{power} = \frac{\sqrt{n_{photons}}}{n_{photons}}. \tag{3.4}$$

Thus, increasing power (number of photons) will grow optical gain faster than shot noise, so shot noise strain decreases with greater power: $h_{shot} \propto \sqrt{1/P}$ (eq. 2.2).

During the first two observing runs, Advanced LIGO ran with 25 W input power. We will double the power during the next run; the expected quantum noise response is shown in fig. 1.

4. 70 W amplifier

The 25 W input power of the previous observing run was the maximum power available from the laser configuration at that time. To double this power, a solid-state amplifier has been installed capable of amplifying to over 70 W of optical power with 4 crystals pumped by fiber-coupled diode lasers. Produced by neoLASE, this 70 W amplifier was specifically designed for minimum-impact integration into the previous table set-up and for quiet operation.

The original Advanced LIGO high power plan was to use a high power oscillator capable of producing 200 W output. During the last observing run, this oscillator was found to inject noise into the interferometer through beam pointing and size jitter. Noise originates from the water system cooling the oscillator crystals; water flow is necessarily turbulent for efficient cooling. Water, brought in from removed chillers, is directed via a breakout manifold under the table and travels to the oscillator in tubes along the optic table. Water flows in a closed loop directly over the oscillator crystals in series with auxiliary cooled dumps, etc. Water motion couples into the beam in two ways:

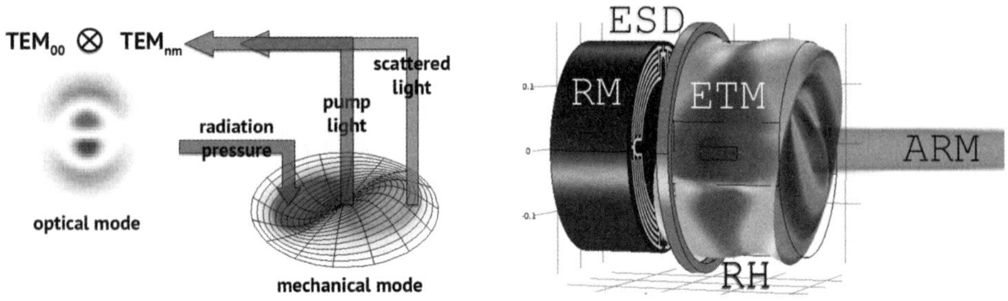

Figure 3. Left: Parametric instability (PI) feedback loop: mechanical motion of the test mass can scatter light into a higher order optical mode which then puts radiation pressure back on the optic, reinforcing the mechanical motion. Right: PI damping components: a) the ring heater (RH) circles the test mass towards the back, altering the radius of curvature and thus the optical mode spacing of the arm cavity b) the electrostatic drive (ESD) on the face of the reaction mass (RM) actuates against the back of the test mass with quadrature control, damping the mechanical modes of the test mass..

- turbulent water flow over the crystals couples directly into the beam
- water flow shaking the optics table couples into optics and thus indirectly into the beam

The 70 W amplifier was designed with minimal water flow needs in mind. It will allow LIGO to double input power for the next run but the more modest (if immediate) power goal allows for the full system to have lower flow requirements. The crystals are cooled indirectly and the system is designed in parallel such that the highest water flow needs no longer dictate the flow rate for all water tubes. Since installation, the 70 W amplifier has proven to output stable power over 70 W and jitter noise from table motion is comparable to that from the previous lower power laser set-up; noise projection is still a factor of 2 below DARM at the worst coupling frequency (fig. 2).

5. Parametric instability

A main problem that arises as LIGO increases power is the growth of parametric instabilities (PI). Long theorized (see Braginsky (2001)), these instabilities were first observed with 12 W input power and will become more problematic with each increase in power. PI have been successfully mitigated during the previous two observing runs, but increased power will require fine tuning of these mitigation schemes.

PI overview Parametric instabilities are run-away mechanisms where the mechanical motion of an arm mirror (test mass) resonance scatters light into a higher-order optical mode in the arm cavity which in turn applies radiation pressure back onto the test mass such that its mechanical motion is reinforced, as shown in Fig. 3.

More specifically, PI can occur when a mechanical mode of an optic falls within the line width of the beat between the carrier and a higher order optical mode (HOM) present in the cavity:

$$f_{mechanical} \sim (p)FSR + (2m+n)TMS \tag{5.1}$$

where p, n, $m \in \mathcal{N}$, $FSR =$ free spectral range $= \frac{c}{2L} = 37.5$ kHz, and $TMS =$ transverse mode spacing ≈ 5.1 kHz, $\nu_0 = \frac{\pi c}{L}$. If this mechanical mode gets rung up (perhaps by some thermal excitation), the oscillating mirror surface will scatter light, effectively modulating the 0,0 mode by $f_{acoustic}$:

$$f_{mod} = f_{0,0} - f_{acoustic} = f_{0,0} - [(p)FSR + (2m+n)TMS] = f_{HOM} \tag{5.2}$$

since TMS is defined as the frequency spacing between 0,0 and HOM, and FSR simply allows for cavity wrapping.

Once 0,0 and HOM are both present in the cavity, the resulting total field applies radiation pressure back on the surface of the optic, closing the loop; if the transverse amplitude distribution of this field overlaps the spacial surface profile of the original acoustic mode, the radiation pressure reinforces the surface motion resulting in an instability which can grow beyond the limits of the cavity control loops.

If we model this mechanism as a classic feedback loop, as shown by Evans (2010), the gain of the loop for a particular mechanical mode m is

$$R_m = \frac{8\pi Q_m P}{M\omega_m^2 c\lambda_0} \sum_{n=1}^{inf} \Re[G_n] B_{m,n}^2 \tag{5.3}$$

where Q_m is the quality factor of the mechanical mode, P is the circulating power, M is the mass of the optic, ω_m is the mechanical frequency, G is the optical transfer function, and B is a term related to the spacial overlap. Note the linear dependence on power.

Controlling PI There are two approaches to mitigating parametric instability:

- Changing the optical gain G_n

Since the light scattered at the mechanical mode frequency into the cavity will only resonant if the frequency is within the line width of the transverse mode spacing, tuning the TMS can decrease this frequency overlap, effectively lowering the loop gain (see Zhao (2005), Gras (2010)). Since

$$TMS = \frac{\nu_0}{\pi} cos^{-1} \sqrt{(1 - L/R_1)(1 - L/R_2)}, \tag{5.4}$$

where $R_{1,2}$ are the radii of curvature of each optic, altering either optic's radius of curvature allows us to tune the optical mode spacing. To do this, LIGO uses ring heaters, annular heating elements that radiate heat onto the outer edge of the test mass; changing the temperature difference across the test mass changes its radius of curvature.

- Changing the quality factor Q_m of the optic

The motion of the test mass can be damped using electrostatic drives (ESD) already in place for low noise actuation (Blair (2017)). Behind each test mass is a reaction mass, whose face has pairs of gold conductors such that a differential voltage can be applied. This puts an electrostatic force on the test mass which couples to the mechanical mode with some coefficient b_m:

$$F_{ESD,m} = b_m A_q \frac{1}{2} (\Delta V)^2 \tag{5.5}$$

There are 4 independent electrostatic drives on each reaction mass, allowing for quadrature longitudinal actuation; by driving equally and oppositely at the mechanical mode frequency in a pattern that roughly matches it's surface profile, the mechanical motion can be damped, reducing the parametric gain below unity to bring the loop back to the stable regime (Miller (2011)).

Both methods have been successfully used to mitigate PIs at both LIGO detectors. However, the first method of tuning to a 'safe zone' of low parametric gain will become obsolete as power is increased: as the parametric gain increases for all mechanical modes linearly with power, areas of previously low gain to which the optical spacing could be tuned before will no longer exist. Additionally, during power up, the temperature differential across the test mass changes, sweeping the optical mode in a short amount of time, leading to large and fast changes in parametric gain. Additionally, mechanical

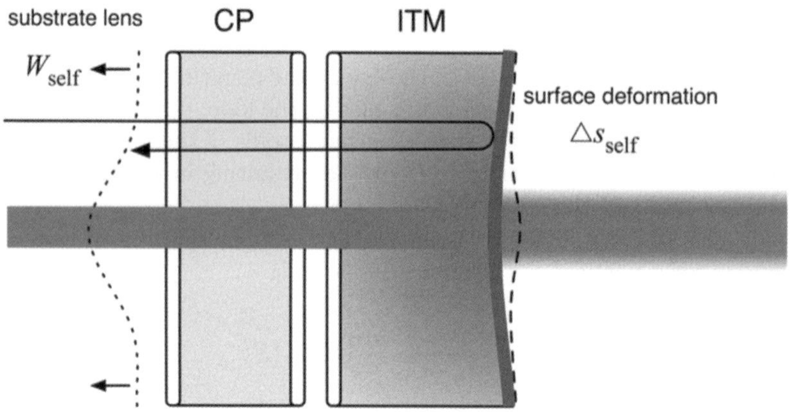

Figure 4. Surface deformation and substrate lensing due to self heating. Here the initial test mass (ITM) forming one of the 4 km arm cavities is shown, whith the blue line indicating the HR surface. Before the ITM hangs the compensation plate (CP), an identical mass sitting between the ITM and the other coupled cavities of the interferometer.

mode frequencies are proportional to Young's modulus E,

$$f_m = A_m \sqrt{E} \Rightarrow \frac{df_m}{dT} = \frac{A_m}{2\sqrt{E}} \frac{dE}{dT} \tag{5.6}$$

so as the optic heats up and as ambient temperatures fluctuate over long lock stretches, the mechanical modes shift in frequency.

6. Thermal effects

High power brings thermal consequences: as optics heat up, thermo-optical distortion occurs (see Brooks (2016)). This is especially relevant in the 4 km arm cavities, where there is large power buildup. Thermal distortion affects mode matching between the coupled cavities of the interferometer, resulting in power losses, control instabilities, and increased noise.

The optics that form the arm cavities have nominal coating absorption of 0.5 ppm. As power in the arms is increased to hundreds of kW, the power absorbed will increase to hundreds of mW. The subsequent temperature differential results in surface deformation of the high-reflectivity side of the test masses and substrate lensing due to the change in index of refraction with temperature of the initial test masses of the arms, as shown in fig. 4. These effects distort the wavefronts of the fields in the interferometer, resulting in poor mode matching.

Thermal compensation system

A thermal compensation system was developed to sense and correct thermally-induced distortions. The system includes:

- Ring heaters

Ring heaters provide deformation of the radius of curvature of the test mass counter to that from self heating. The arm cavity beam, roughly centered on the test mass, heats the center of the optic, effectively increasing the radius of curvature; the ring heater, in contrast, heats the outer radius of the optic, effectively decreasing the radius of curvature. Ring heaters have large thermal time constants so are tuned for the steady state operation of the locked interferometer.

- CO_2 lasers

One more degree of control is necessary for mode matching. CO_2 lasers are projected onto the back of the compensation plate (CP), creating a tunable lens between the arm cavities and the other cavities of the interferometer. Before impinging on the CP, the beam passes through a mask that sets a spacial distribution; this mask can be interchanged, allowing for tunable beam shape. The nominal design consists of central heating and annular heating, the former to counteract strong arm cavity central heating transients during locklosses and the latter to tune the residual distortion from the combined ring heater and central heating settings.
All concepts are illustrated together in fig. 4.

7. Conclusion

Higher power during the next observing run will give LIGO better sensitivity at high frequency by decreasing shot noise, resulting in higher SNR for gravitational wave detections over a larger frequency band. The increase of power will come from a newly installed, quiet 70 W amplifier and will correspond with squeezer installation. With a power increase comes parametric instabilities and thermal detuning. Control schemes have been developed and successfully tested for avoiding and actively damping PI. Hardware is in place to counteract thermal effects and is being commissioned for the higher power requirements.

References

Saulson, P. *Fundamentals of Interferometric Gravitational Wave Detectors.* World Scientific Publishing Co. 2017.
Braginsky, V. B., Strigin, S. E., & Vyatchanin, S. P. 2001, *Phys. Lett.* A287, 331-338
Zhao, C., *et al.* 2005, *Phys. Rev. Lett.* 94, 121102
Blair, C., *et al.* 2017, *Phys. Rev. Lett.* 118, 151102
Gras S., *et al.* 2010, *Classical and Quantum Gravity* 27, 20
Evans, M., Barsotti, L., & Fritschel, P. 2010, *Phys. Lett.* A374, 665-671
Miller, J. 2011, *Phys. Lett.* A375, 788-794
Brooks, A., *et al.* 2016, *Appl. Opt.* 55, 8256-8265
Brooks, A., Kelly, T., Veitch, P., & Munch, J. 2007, *Opt. Express* 15, 10370-10375

Gravitational Wave Astrophysics: Early Results from Gravitational Wave Searches and Electromagnetic Counterparts
Proceedings IAU Symposium No. 338, 2017
G. González & R. Hynes, eds.
doi:10.1017/S1743921318003757

Strong-lensing of Gravitational Waves by Galaxy Clusters

Graham P. Smith,[1†] Christopher Berry,[1,2] Matteo Bianconi,[1] Will M. Farr,[1,2] Mathilde Jauzac,[3,4,5,6] Richard Massey,[3] Johan Richard,[7] Andrew Robertson,[4] Keren Sharon,[8] Alberto Vecchio,[1,2] and John Veitch.[9]

[1]School of Physics and Astronomy, University of Birmingham, Birmingham, B15 2TT, England
[2]Birmingham Institute of Gravitational Wave Astronomy, University of Birmingham, Birmingham, B15 2TT, England
[3]Centre for Extragalactic Astronomy, Department of Physics, Durham University, Durham DH1 3LE, England
[4]Institute for Computational Cosmology, Durham University, South Road, Durham DH1 3LE, England
[5]Astrophysics and Cosmology Research Unit, School of Mathematical Sciences, University of KwaZulu-Natal, Durban 4041, South Africa
[6]Laboratoire d'Astrophysique École Polytechnique Fédérale de Lausanne (EPFL), Observatoire de Sauverny CH-1290 Versoix, Switzerland
[7]CRAL, Observatoire de Lyon, Université Lyon 1, 9 Avenue Ch. André, 69561 Saint Genis Laval Cedex, France
[8]Department of Astronomy, University of Michigan, 500 Church St., Ann Arbor, MI 48109 USA
[9]School of Physics and Astronomy, University of Glasgow, G12 8QQ, Scotland
[†]Email: gps@star.sr.bham.ac.uk

Abstract. Discovery of strongly-lensed gravitational wave (GW) sources will unveil binary compact objects at higher redshifts and lower intrinsic luminosities than is possible without lensing. Such systems will yield unprecedented constraints on the mass distribution in galaxy clusters, measurements of the polarization of GWs, tests of General Relativity, and constraints on the Hubble parameter. Excited by these prospects, and intrigued by the presence of so-called "heavy black holes" in the early detections by LIGO-Virgo, we commenced a search for strongly-lensed GWs and possible electromagnetic counterparts in the latter stages of the second LIGO observing run (O2). Here, we summarise our calculation of the detection rate of strongly-lensed GWs, describe our review of BBH detections from O1, outline our observing strategy in O2, summarize our follow-up observations of GW170814, and discuss the future prospects of detection.

Keywords. gravitational lensing, gravitational waves, galaxies: clusters: individual (MACS J0140.0−0555, MACS J1311.0−0310, RCS 0224−0002, Abell 3084, SMACS J0304.3−4401)

---

1. Strong-lensing of gravitational waves by galaxy clusters

The detection of strongly-lensed gravitational wave (GW) sources will create opportunities to explore binary compact objects at higher redshifts and/or that are intrinsically fainter than those that are detectable without assistance from lensing. This greater "reach" of GW detectors that can be achieved with lensing is due to the gravitational magnification, μ, that boosts the amplitude of the strain signal (A) at an interferometer by a factor $\sqrt{\mu}$ and the spectral flux density of any accompanying electromagnetic signal by a factor μ. In addition to increasing the sensitivity of observations, the gravitational magnification has important consequences for the interpretation of the strain and EM signals. In particular, there is a degeneracy between the luminosity distance (D_L) to the source and the gravitational magnification, such that $A \propto D_L^{-1}|\mu|^{0.5}$. Therefore, if it is assumed that $\mu = 1$ for a source that is actually lensed, then D_L and the redshift of the source (z) will be under-estimated. Moreover, given that the rest-frame mass of the source goes as $(1+z)^{-1}$, the mass of the source would be over-estimated if $\mu = 1$ is erroneously assumed (Wang, Stebbins & Turner 1996).

The early detections of binary black hole (BBH) GW sources by LIGO and Virgo are intriguing in the context of gravitational lensing because a large fraction of the reported BHs have rest frame masses larger than the most massive stellar mass BHs detected thus far (Abbott *et al.* 2016a, 2016b, 2017a, 2017b; Farr *et al.* 2011). This encourages speculation that one or more of the sources could have been gravitationally lensed.

The realisation that a GW source is strongly-lensed would also open up the possibility to detect the same GW source on more than one occasion by virtue of the different arrival times along multiple sight lines that connect an observer with a strongly-lensed source. Given the sub-millisecond accuracy of the measurement of the arrival time of a GW signal, it would be possible to measure the time delay between different sight lines through a galaxy cluster to a single GW source to a similar accuracy. Such accuracy is $\gtrsim 8$ orders of magnitude better than is achievable with multiply-imaged supernovae and quasars (Fohlmeister *et al.*, 2007; Rodney *et al.*, 2016). Discovery of one or more multiply-imaged GW sources would therefore deliver an unprecedented constraint on the local mass distribution in cluster lenses, and potentially offer a new and accurate probe of the Hubble parameter (e.g. Liao *et al.* 2017). Moreover, multiple detections of the same GW are unlikely to be made when the detectors are in the same orientations relative to the source. Therefore, observations of a multiply-imaged GW source should help to further constrain GW polarizations and potentially achieve new tests of General Relativity, since we would effectively observe the same signal with a greater number of detectors (Chatziioannou, Yunes & Cornish 2012).

2. The probability of strong-lensing

Given that the BBH GW sources are located at $z \sim 0.1 - 0.3$ (assuming $\mu = 1$), a magnification factor of $\mu \sim 30 - 300$ is required to reinterpret the strain signal as originating from a source at (say) $z \simeq 1$ (Smith *et al.* 2018). The physical region from which GWs emerge is ~ 100 km in size. It is therefore possible for a GW source to be very closely aligned with the caustic of a gravitational lens, and thus achieve such a high magnification value (Ng *et al.* 2018) – i.e. we can treat GWs as point sources. In contrast, galaxies have typical physical sizes of $\sim 1 - 10$ kpc, and therefore cannot be treated as point sources at optical wavelengths. Hilbert *et al.* (2008) predicted that the optical depth to strong-lensing of point sources is dominated by galaxy clusters, with the optical depth peaking at $M \simeq 10^{14} M_\odot$. This peak, and the dominance of clusters, is predicted to be more pronounced for large lens magnifications, i.e. $\mu > 10$ (Figure 5 of Hilbert *et al.*). These predictions appear to be borne out by observations, in that the typical lens magnifications suffered by quasars – a class of point source, and thus directly relevant to lensing of GWs – that are strongly-lensed by clusters is typically $\mu > 10$ (Oguri *et al.* 2010, 2013; Sharon *et al.* 2017), whilst the magnification of quasars strongly-lensed by galaxies is typically $\mu < 10$ (Inada *et al.* 2014). Therefore, the available predictions and observations both point to galaxy clusters being more relevant than galaxies (Broadhurst *et al.* 2018; Li *et al.* 2018; Ng *et al.* 2018) to the discovery of strongly-lensed GWs with LIGO-Virgo.

There are currently 130 known and spectroscopically confirmed cluster strong lenses (Smith *et al.* 2018 and references therein). It is important to stress that spectroscopic confirmation of the redshift of multiply-imaged galaxies seen through these cluster cores is essential to achieve robust constraints on the gravitational optics (e.g. Smith *et al.* 2009). Most of the cluster lenses are at redshifts of $z \sim 0.2-0.5$; at lower redshifts strong-lensing clusters become increasingly rare due to the small observable cosmic volume, and higher redshifts await systematic exploration.

The assumption of $\mu = 1$ in the LIGO-Virgo analysis of BBH detections in the first and second observing runs (O1 and O2) rightly reflects the fact that the probability of strong-lensing is small. This is due to a combination of factors including the large magnifications noted above, and existing constraints on the local BBH merger rate. Nevertheless, Smith *et al.* (2018) showed that the probability is non-zero. Based on the known population of cluster lenses and the GW detections from O1 (Abbott *et al.*, 2016b), they estimated a lower limit on the rate of detectable strongly-lensed GWs of $R_{\rm detect} \simeq 10^{-5}$ per detector year. This corresponds to rejecting the hypothesis that one of the O1 GW detections was multiply-imaged at $\lesssim 4\sigma$.

3. Known strong-lensing clusters and LIGO's O1 BBHs

Smith *et al.* (2018) identified three strong-lensing clusters with celestial coordinates consistent with the 90% credible sky localisations of GW sources detected in LIGO's O1 (Abbott *et al.*, 2016b), of which none are consistent with GW150914, two are consistent with GW151226 (MACS J0140.0−0555 and MACS J1311.0−0310) and one is consistent with LVT151012 (RCS 0224−0002). Following Jauzac *et al.* (2016) we used detailed parameterized models of the mass distribution in each of these cluster lenses to explore whether and when another appearance of GW151226 and LVT151012 would be detectable, if they have been strongly-lensed. In summary, the next appearance will be detectable at the current LIGO-Virgo sensitivity in roughly half of the cases, and on a timescale of up to 3 years from the initial detection. The "re-detection rate" of $\sim 50\%$ arises because the relevant strong-lensing configurations typically involve one caustic crossing, and thus produce three images, of which two are very highly magnified and detectable and the third is not. Therefore, if the initial GW detection corresponds to the first of the two highly magnified images, then the second of the two highly magnified images is generally detectable. If the initial GW detection corresponds to the second of two highly magnified images, then the next appearance corresponds to the third much less strongly magnified and thus undetectable image.

4. Gravitationally Lensed Gravitational Wave Hunters and O2

The Gravitationally-lensed Gravitational Wave Hunters collaboration began observing known strong-lensing clusters located within the 90% credible sky localisations of new GW detections in August 2017. The aim of these observations was to detect an EM counterpart to a GW based on the speculation that it had been strongly-lensed by a cluster. We used the GMOS instrument on Gemini-South and the MUSE instrument on VLT to examine the strong-lensing regions of clusters down to depths of AB $\simeq 25$ per visit. For a nominal gravitational magnification of $|\mu| = 100$, we therefore in principle reach a sensitivity of AB $\simeq 30$ in the search for an EM counterpart to GWs. This is the most sensitive search conducted to date. The first visit with each telescope took place as soon as possible after the LIGO-Virgo alert via a rapid target of opportunity observation, and subsequent visits took place up to ~ 1 week later.

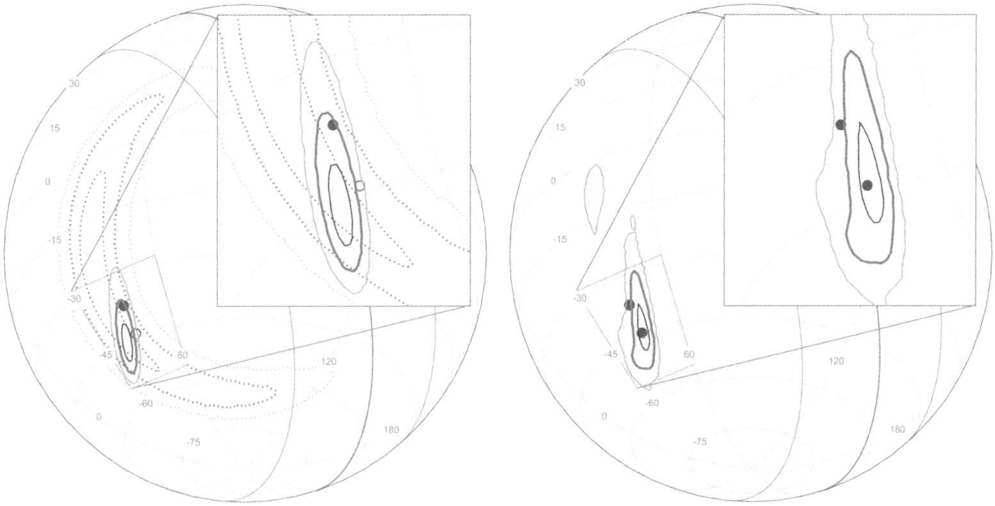

Figure 1. LEFT – Initial BAYESTAR skymap for GW170814 from GCN21474, showing the LIGO only contours (dotted), joint LIGO-Virgo contours (solid), Abell 3084 (filled blue), and SMACS J0304.3−4401 (open blue). RIGHT – Initial LALInference skymap from GCN21493, showing updated sky localisation based on LIGO-Virgo data (solid) and the locations of both clusters. SMACS J0304.3−4401 is closer to the peak of the sky localisation than Abell 3084 in the revised map. In both panels the 90% credible region is shown as the thick solid contour, and galactic latitudes of ±20° are shown in red.

We observed two strong-lensing clusters associated with the sky localisation of GW170814 (Abbott *et al.* 2017b), namely Abell 3084 and SMACS J0304.3−4401 (Smith *et al.* 2017). Figure 1 shows the location of the clusters and the LIGO-Virgo credible regions on the celestial sphere. The small change in the sky localisation between the initial BAYESTAR and preliniary LALInference skymaps, and the relationship between these and the cluster coordinates explains why we initially observed Abell 3084 and subsequently switched to SMACS J0304.3−4401.

5. Future prospects

We consider the timescale on which the rate of detection of strongly-lensed GWs will approach $R_{\rm detect} \simeq 1$ per detector year. Two factors contribute to this timescale: (1) gains in sensitivity of the LIGO-Virgo detectors, and (2) expansion of the sample of known strong-lensing clusters. LIGO is scheduled to reach design sensitivity in 2020, by which time it will probe a $\sim 10\times$ larger volume than was the case in O1 and O2 (Abbott *et al.*, 2017c). This improved sensitivity will lead to less extreme magnifications being required to reinterpret GW signals as coming from redshifts higher than the cluster lenses. We estimate that the greater sensitivity and consequent reduced magnification requirement will together increase the rate to $R_{\rm detect} \simeq 10^{-2}$ per detector year, based on the same population of 130 known strong lensing clusters discussed above.

Clearly the incidence of strong-lensing of GWs is independent of the completeness of the available sample of strong-lensing clusters. However, ability to recognise GWs as being strongly-lensed does depend on knowledge of the lens population. Therefore $R_{\rm detect}$ will grow as more strong-lensing clusters are discovered. Forecasts for upcoming large-scale optical/near-infrared surveys indicate that *Euclid* and LSST will find of order 1 strong-lensing cluster per square degree in the redshift range $z \sim 0.2 - 0.5$ considered above

(Boldrin *et al.*, 2016). It should therefore be possible to achieve a further two orders of magnitude gain to give $R_{\rm detect} \simeq 1$ per detector year during the 2020s. Precisely when this will be achieved will depend on the observing strategy of each survey, and the efficiency of strong-lensing cluster detection in the survey data.

Additional GW detectors, KAGRA and LIGO-India, are planned to begin operation in the period 2023-2025 (Abbott *et al.*, 2017c), i.e. in parallel with *Euclid*/LSST discovery of strong-lensing clusters. These detectors will increase the volume within which GWs can be detected by a further factor of 3 beyond the LIGO design sensitivity. This will help to increase the rate of strongly-lensed GW detections in the mid-2020s.

Acknowledgments

GPS thanks Iair Arcavi for assistance with producing Figure 1. We are grateful for awards of observing time by the Directors of the European Southern and Gemini Observatories. We acknowledge financial support from the Science and Technology Facilities Council and the European Research Council.

References

Abbott B. P., *et al.*, 2016a, *Astrophys. J. Lett.*, 818, 22
Abbott B. P., *et al.*, 2016b, *Phys. Rev. X*, 6, 041015
Abbott B. P., *et al.*, 2017a, *Phys. Rev. Lett.*, 118, 221101
Abbott B. P., *et al.*, 2017b, *Phys. Rev. Lett.*, 119, 141101
Abbott B. P., *et al.*, 2017c, arXiv:1304.0670v4
Boldrin M., *et al.*, 2016, *Mon. Not. R. Astron. Soc.*, 457, 2738
Broadhurst T., *et al.*, 2018, arXiv:1802.05273
Chatziioannou K., Yunes N., & Cornish N., 2012, *Phys. Rev. D*, 86, 022004
Farr W. M., *et al.*, 2011, *Astrophys. J.*, 741, 103
Fohlmeister J., *et al.*, 2007, *Astrophys. J.*, 662, 62
Hilbert S., *et al.*, 2008, *Mon. Not. R. Astron. Soc.*, 386, 1845
Inada N., *et al.*, 2014, *Astron. J.*, 147, 153
Jauzac M., *et al.*, 2016, *Mon. Not. R. Astron. Soc.*, 457, 2029
Li S.-S., *et al.*, 2018, arXiv:1802.05089
Liao K., *et al.*, 2017, *Nature Communications*, 8, 1148
LIGO Scientific Collaboration and Virgo Collaboration, 2017, GCN Circ. 21474, https://gcn.gsfc.nasa.gov/gcn3/21474.gcn3
LIGO Scientific Collaboration and Virgo Collaboration, 2017, GCN Circ. 21493, https://gcn.gsfc.nasa.gov/gcn3/21493.gcn3
Ng K. K. J., *et al.*, 2018, *Phys. Rev. D*, 97, 023012
Oguri M., *et al.*, 2010, *Pub. Astron. Soc. Japan*, 62, 1017
Oguri M., *et al.*, 2013, *Mon. Not. R. Astron. Soc.*, 429, 482
Rodney S. A., *et al.*, 2016, *Astrophys. J.*, 820, 50
Sharon K., *et al.*, 2017, *Astrophys. J.*, 835, 5
Smith G. P., *et al.*, 2009, *Astrophys. J. Lett.*, 707, 163
Smith G. P., *et al.*, 2017, GCN Circ. 21692, https://gcn.gsfc.nasa.gov/gcn3/21692.gcn3
Smith G. P., *et al.*, 2018, *Mon. Not. R. Astron. Soc.*, 475, 3823
Wang Y., Stebbins A., & Turner E. L., 1996, *Phys. Rev. Lett.*, 77, 2875

Gravitational Wave Astrophysics: Early Results from Gravitational Wave Searches and Electromagnetic Counterparts
Proceedings IAU Symposium No. 338, 2017
G. González & R. Hynes, eds.

doi:10.1017/S1743921318003575

How Einstein's Theory of Relativity Gives us $E = mc^2$ and the Atomic Bomb

Richard Conn Henry

Henry A. Rowland Department of Physics & Astronomy,
The Johns Hopkins University, 3400 North Charles St, Baltimore, Maryland, USA
email: `henry@jhu.edu`

Abstract. The LIGO *direct-detection* of gravitational waves arriving from cosmic sources—now, happily including (as of this meeting!) the merging of two neutron stars—opens a new chapter in our understanding of physics itself: for General Relativity, conceptually so extremely simple, has robustly produced predictions that have invariably been found to be correct when tested. My poster (page 3 of this paper) is intended for high school students who have just learned simple algebra. My derivation of the famous $E = mc^2$ from the Pythagorean Theorem necessarily requires an algebraic expansion that is due to Newton, but apart from that it is simplicity itself: a transparent introduction to what all of physics is today: the construction of mathematics that, miraculously, reproduces our observations of the world—and which also successfully predicts the results of future observations—as so magnificently demonstrated at this glorious Symposium!

Keywords. Pythagoras, Einstein, Minkowski, relativity, education

---

1. Introduction

My education poster is the third page of this paper—the rest guide educators in their use of the poster. For the poster is terse—it is the job of the teacher to guide the student and induce her to simply write out the entire poster! For that is the secret of coping with equations. For me, just as for (I think) the typical student, equations, at first, seem simply to bounce off the eyeballs. But, if each equation is actually written out, then the miracle occurs: for to be written out requires each equation to have passed through the eyes; and into the brain; and then, out onto the page—and understanding occurs.

Good students are superb *readers:* just a glance at words in a text conveys meaning. But a glance at an equation is another story! Writing out an equation produces *understanding:* and here, the algebra is trivial, and no steps, however simple, have been left to the reader.

2. The Pythagorean Theorem

The Pythagorean Theorem is known to every educated person, not just in its lyrical statement that "the square on the hypotenuse is equal to the sum of the squares on the other two sides," but also in its algebraic expression, $s^2 = x^2 + y^2$, which of course is what we will need when we will—momentarily—zoom to Einstein's special relativity.

Look at the poster (page 3 of this paper) and see that the first step is the Pythagorean Theorem, here illustrated in such a way that the concept of invariance under a transformation is made clear. The transformation is rotation of the Cartesian coordinates, which clearly should not, and clearly does not, change the length of the hypotenuse.

My poster needs no calculus: simple high school algebra is all that is needed. But I do use one *notation* from calculus: if x, say, is a coordinate, then an interval along the x axis (e.g. $x_2 - x_1$) I denote by $dx \equiv x_2 - x_1$. (While in calculus dx would mean an *infinitesimal* interval in the x direction, in this paper dx means *any* interval in the x direction.)

I now introduce a *supplement* to my poster to assist the teacher who is guiding her students through the poster. The teacher should ask the student, "how do we know that the Pythagorean Theorem itself is actually true?"

The teacher should then show the student **Figure 1**, which is an instantaneous and (most importantly!) *non-algebraic* proof of the Pythagorean Theorem, revealing that theorem to be a *property of the space in which we live.*

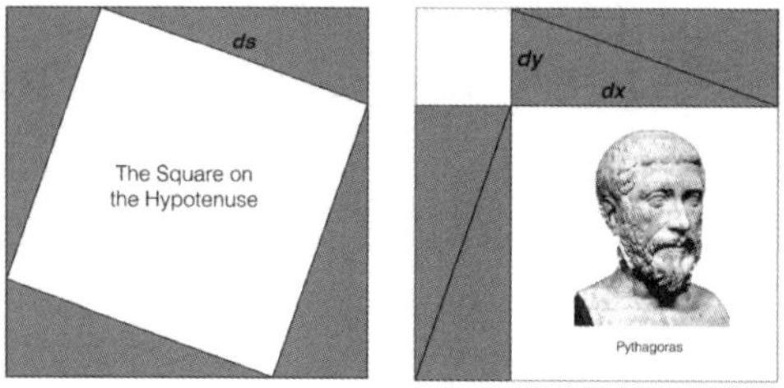

Figure 1. The Theorem of Pythagoras: the two big squares are identical in area. All 8 triangles are identical in area. Equals taken from equals leaves equals, so if you throw out all of the triangles, what remains in each of the two diagrams must be equal. (And: no algebra is used at all!)

Now, while our figure does include identification of dx, dy, and ds (the hypotenuse), these are *not* involved in the proof. So when the algebra *is* introduced, it *reproduces* reality! That is why algebra was invented. That is also why it is so powerful! Now we, every one of us, are stuck, forever, at the present moment of time. But algebra allows us, intellectually and (it turns out) correctly to *navigate* time—we can extend algebra to predict additional aspects of reality!

3. William Rowan Hamilton

Hamilton, living after Newton but before Einstein, was perhaps the best mathematical physicist who ever lived. But being good at mathematics is not everything, as we shall see! For Hamilton asked a question which perhaps has occurred to your students: the Pythagorean Theorem in three dimensions is clearly $s^2 = x^2 + y^2 + z^2$, and of course, beyond that, it is easy enough to think of time as being a fourth dimension. And so Hamilton examined $s^2 = x^2 + y^2 + z^2 + t^2$ which is what the Pythagorean Theorem would be in four dimensions. But that would be only adding in a fourth (and non-existent) *space* dimension—despite our optimistically labeling it *t*ime. So—Hamilton gave up!

Well, just three years after Einstein's 1905 discovery of his famous theory of relativity, his former teacher, Hermann Minkowski, discovered that Einstein's hugely successful "Theory of Relativity" was nothing more that the discovery that in our universe, the separation of two events (for example, two snaps of your fingers, as you wave your arm about) is (brace yourself) $s^2 = x^2 + y^2 + z^2 - t^2$. This realization, by Minkowski, is the most important insight into the nature of the universe in the history of human thought.

And surely poor Hamilton rolled over in his grave: we now see that having imagination is at least as important as mathematical ability in the exploration of the universe!

4. The Statement of Einstein's Theory of Relativity

$d\alpha^2 + d\beta^2 + d\gamma^2 - d\tau^2 = dx^2 + dy^2 + dz^2 - dt^2$ is Einstein's theory of relativity, complete (although Einstein himself did not realize that in 1905 when he seeded its discovery!)

The invariance is under a transformation that is quite different from the transformation (rotation) that is illustrated at the top of the poster. No rotation is involved in the transformation of special relativity—just a transformation from a *stationary* coordinate system to coordinate system that is *moving* (with respect to the first) in a fixed direction and with constant velocity (say, $v = dx/dt$). The simplest transformation imaginable!

IAU Symposium 338 Gravitational Wave Astrophysics Baton Rouge, Louisiana

How Einstein's Theory of Relativity gives us $E = mc^2$ and the Atomic Bomb

The first step:

We remind ourselves about Pythagorean invariance under a two-dimensional rotation in space:

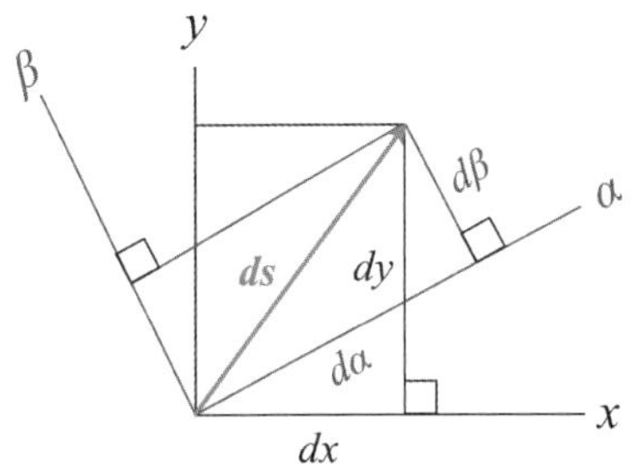

$$d\alpha^2 + d\beta^2 = ds^2 = dx^2 + dy^2$$

And then—the crucial next step:

Suppose that your sister is moving along your x-axis at a constant speed v

and that during that motion, your sister snaps her fingers twice, $d\tau$ seconds apart —

for you, those two finger-snaps are separated in space by dx and are separated in time by dt

---

RELATIVITY: → **Pythagoras, Einstein, and Minkowski assert that $d\tau$ and dt are related by:**

your Sister → $d\alpha^2 + d\beta^2 + d\gamma^2 - d\tau^2 = ds^2 = dx^2 + dy^2 + dz^2 - dt^2$ **← You**

If so, then $0^2 + 0^2 + 0^2 - d\tau^2 = dx^2 - 0^2 - 0^2 - dt^2$. **And now—let us process this result!**

$d\tau^2 = dt^2 - dx^2$ ← If they are right (and experiment shows they *are*) **time slows down** for your sister!

but $dx = v\,dt$ so $d\tau^2 = (1 - v^2)\,dt^2$ and so $(\frac{dx}{dt} =)\ v \leqslant \mathbf{1}$! ← A universal **speed-limit** c **= ONE**

For the case $v = 1,\ d\tau = 0$ ← and so *photons never exist!* Next: $\mathbf{E} \equiv$ **Energy** and $\mathbf{p} \equiv$ **momentum**

$d\tau^2 = dt^2 - dx^2 \rightarrow 1 = (\frac{dt}{d\tau})^2 - (\frac{dx}{d\tau})^2 = \frac{1}{1-v^2} - (\frac{dx}{dt} \cdot \frac{dt}{d\tau} = v\frac{dt}{d\tau})^2$ **or** $\mathbf{1} = \frac{1}{1-v^2} - v^2(\frac{1}{1-v^2})$ **RELATIVITY**

Now multiply through by the square of mass m: $m^2 = m^2(\frac{1}{1-v^2}) - m^2v^2(\frac{1}{1-v^2}) \equiv \mathbf{E^2} - \mathbf{p^2}$

$\gamma \equiv \sqrt{\frac{1}{1-v^2}} = 1 + \frac{1}{2}v^2 + \frac{3}{8}v^4 + \ldots$ ← from **Newton** [for $v < 0.1,\ 1 \leqslant \gamma < 1.00503\cdots$ i.e. $\gamma \approx \mathbf{1}$]

and so: $m^2 = m^2\gamma^2 - (\gamma m v)^2 = m^2(1 + \frac{1}{2}v^2 + \frac{3}{8}v^4 + \cdots)^2 - (\gamma m v)^2$ $\quad \mathbf{E} = \frac{m}{\sqrt{1-v^2}} \quad \mathbf{p} = \frac{m\,v}{\sqrt{1-v^2}}$

$m^2 = m^2[\mathbf{1} + \frac{1}{2}(\frac{v}{c})^2 + \frac{3}{8}(\frac{v}{c})^4 + \cdots]^2 - m^2[\gamma(\frac{v}{c})]^2$ $\quad c = 1$ *or –but only by convention!–* $= 299{,}792{,}458\ m/s$

$m^2c^2 = \frac{1}{c^2}[m\,c^2 + \frac{1}{2}m\,v^2 + \frac{3}{8}m\,\frac{v^4}{c^2} + \ldots]^2 - [\gamma\,m\,v]^2$ ← **we spot** Newton's $E = \frac{1}{2}m\,v^2$ **and** his $p = m\,v$!

$\boldsymbol{m^2\,c^2} = \frac{\mathbf{E^2}}{\mathbf{c^2}} - \boldsymbol{p^2}$ and so, **beyond Newton: →** $\mathbf{E} = \mathbf{m\,c^2} + \frac{1}{2}m\,v^2 + \cdots$ and $\mathbf{p} = \gamma\,m\,v$

And if $v = 0$ we find that—from the (extended) Pythagorean theorem—we have successfully derived the most famous equation in human history: $E = m\,c^2$ ... which has given us the Atomic Bomb:

$neutron + {}^{235}U \rightarrow {}^{236}U \rightarrow {}^{92}Kr + {}^{141}Ba + \langle E = [m_U - (m_{Kr} + m_{Ba})] \times c^2 \rangle + 3\,neutrons$ out →

Maryland Space Grant Consortium henry@jhu.edu

5. Consequences of Einstein's Theory of Relativity

The poster digs into this "extended Pythagorean Theorem" (which, today, is called Einstein's Theory of Relativity): with the help of Newton, we will understand its *meaning.*

Suppose that you are stationary in your coordinate system (x, y, z, t) [t being the time according to your (stationary) clocks] — and *what happens* is that your sister, who is moving with respect to you in a straight line (with constant velocity v) in her moving coordinate system $(\alpha, \beta, \gamma, \tau)$ [with τ being the time as measured by all clocks that are moving with her in her (moving) reference frame], *snaps her fingers twice:* two "events!"

Let your sister's (moving) α axis lie atop your (stationary) x axis, and let the motion take place along that axis. Then we can simply ignore the other two space dimensions.

In your sister's moving reference frame the two finger snaps occur where she is: at locations $(0, 0, 0, \tau_1)$ and $(0, 0, 0, \tau_2)$—and of course have no spatial separation, only a time separation $\tau_2 - \tau_1 \equiv d\tau$; while for you, her two finger snaps have locations$(x_1, 0, 0, t_1)$ and $(x_2, 0, 0, t_2)$, and so have a spacetime separation of $(x_2 - x_1, 0, 0, t_2 - t_1) \equiv (dx, 0, 0, dt)$.

So our relativity equation, for the separation of her two finger-snaps, is $0^2 + 0^2 + 0^2 - d\tau^2 = dx^2 + 0^2 + 0^2 - dt^2$, or simply $d\tau^2 = dt^2 - dx^2$. But the distance you see your sister go in dt is $dx = v\,dt$ and so we get the poster result that $d\tau^2 = (1 - v^2)\,dt^2$. Now any number, positive OR negative, when squared, is positive. But if v were greater than one, the right-hand side of our equation would be negative—and yet required to be equal to the left-hand side which is necessarily positive! We conclude that if Einstein's theory governs the universe in which we live, **velocities greater than 1 are impossible.**

Experiment has shown that this is true. The fastest speed we have ever measured is the speed of light, which is: **one** light-year per year. That is, **1**. No physical units!

[Ah, units! One reason physics is so confusing is is that we humans are country bumpkins: the speed of light is not really $c = 299{,}792{,}458$ meters per second, it is $c = 1$. One! **1**! No units! Put $E = m$ on your T-shirt, not $E = m\,c^2$! [And Planck's constant (quantum mechanics) is not 6.625×10^{-27} erg-seconds: **it is**: yes you guessed it, **1**!] We use the units that we do partly so that we won't have to use extremely large or extremely small numbers in our daily lives. Our daily work takes place in an extremely limited portion of the available universe. In fact we are all hicks in an *inconceivably big* universe that contains things that are *inconceivably small.* For example, inside a billion-solar-mass black hole is a singularity, having that mass, that is much smaller than one single proton!]

More simple processing yields my poster's $m^2 c^2 = \frac{1}{c^2}(m\,c^2 + \frac{1}{2}m\,v^2 + \ldots)^2 - (\gamma\,m\,v)^2$. If a motionless mass m is in front of you the equation says that $m = m$ which is true!

But if you enter a different frame, in which the mass moves at v, our equation says that what is present is not m but is rather the (measurable) quantities $(E \equiv)\,m\,c^2 + \frac{1}{2}m\,v^2 + \ldots$ and $(p \equiv)\,\gamma\,m\,v$, where we can now (joyfully!) recognize Newtonian kinetic energy and Newtonian momentum hiding inside Einstein's results, and we have discovered that with the mass in motion $E = m$ plus, if $v \ll 1$, just a (relatively) *tiny bit more* energy!

(By the way: I hope you noticed that our equation $1 = \frac{1}{1-v^2} - v^2\,(\frac{1}{1-v^2})$ is an identity!)

A teacher can bring home the importance of all this to her students by pointing out that LHC (Europe's Large Hadron Collider) pumps arbitrarily large amounts of energy into single particles yet despite their huge energy none of those particles ever exceeds $v = c = 1$*!* So *where does their extra kinetic energy reside?* It is in the infinite number of additional terms beyond $\frac{1}{2}m\,v^2$ that I have designated by simply $+\ldots$

I thank Professor Gabriela González for creating this timely Symposium.

Reference

Henry, R. C. 2005, *Nature*, 436, 29

Author index

Annis, J. – 61, 65, 72

Baliyan, K. S. – 29
Berdyugin, A. – 29
Berry, C. – 98
Bianconi, M. – 98
Butler, R. E. – 61

Cao, L. – 14
Caton, D. B. – 29
Chatziioannou, K. – 22
Ciprini, S. – 29

Dey, L. – 29
Díaz, M. C. – 80

Farr, W. M. – 98

Gazeas, K. – 29
Gerosa, D. – 22
Gezari, S. – 46
Gopakumar, A. – 29
Goźlińska, B. – 37

Haque, S. – 29
Hardwick, T. – 90
Haster, C.-J. – 22
Henry, R. C. – 103
Herner, K. – 61, 65
Hudec, R. – 29

Jauzac, M. – 98
Jermak, H. – 29

Kennea, J. A. – 53
Kidger, M. – 29
Kobayashi, S. – 1

Lamb, G. P. – 1
Lambas, D. G. – 80
Lamberts, A. – 40
Liu, T. – 46
Lu, Y. – 14

Macri, L. M. – 80
Massey, R. – 98
Matsumoto, K. – 29
Mauri, J. – 29
Mikkola, S. – 29

Nilsson, K. – 29

Piirola, V. – 29

Reichart, D. E. – 29
Richard, J. – 98
Robertson, A. – 98

Sadakane, K. – 29
Sharon, K. – 98
Smith, G. P. – 98
Soares-Santos, M. – 61, 65

Tavani, M. – 84
Tohuvavohu, A. – 53

Vecchio, A. – 98
Veitch, J. – 98
Verrecchia, F. – 84
Vitale, S. – 22

Yang, S. – 9

Zadrożny, A. – 37
Zhao, Y. – 14
Zimmerman, A. – 22
Zola, S. – 29